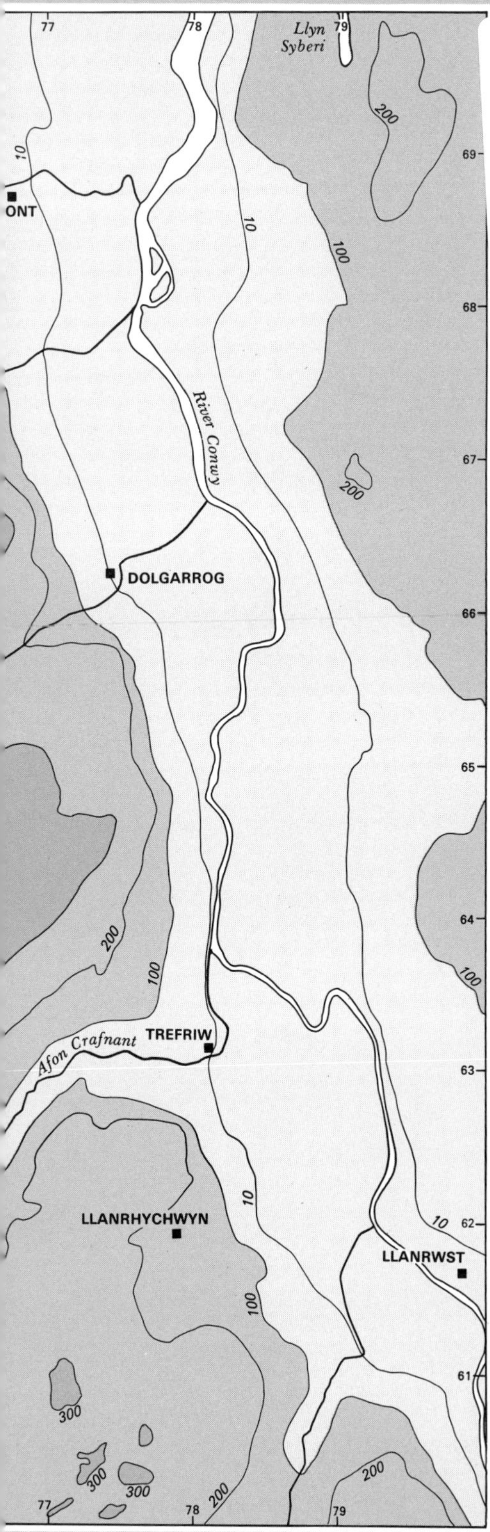

	100-200 metres
	400-500 metres
	300-400 metres
	600-700 metres
	500-600 metres
	800-900 metres
	700-800 metres

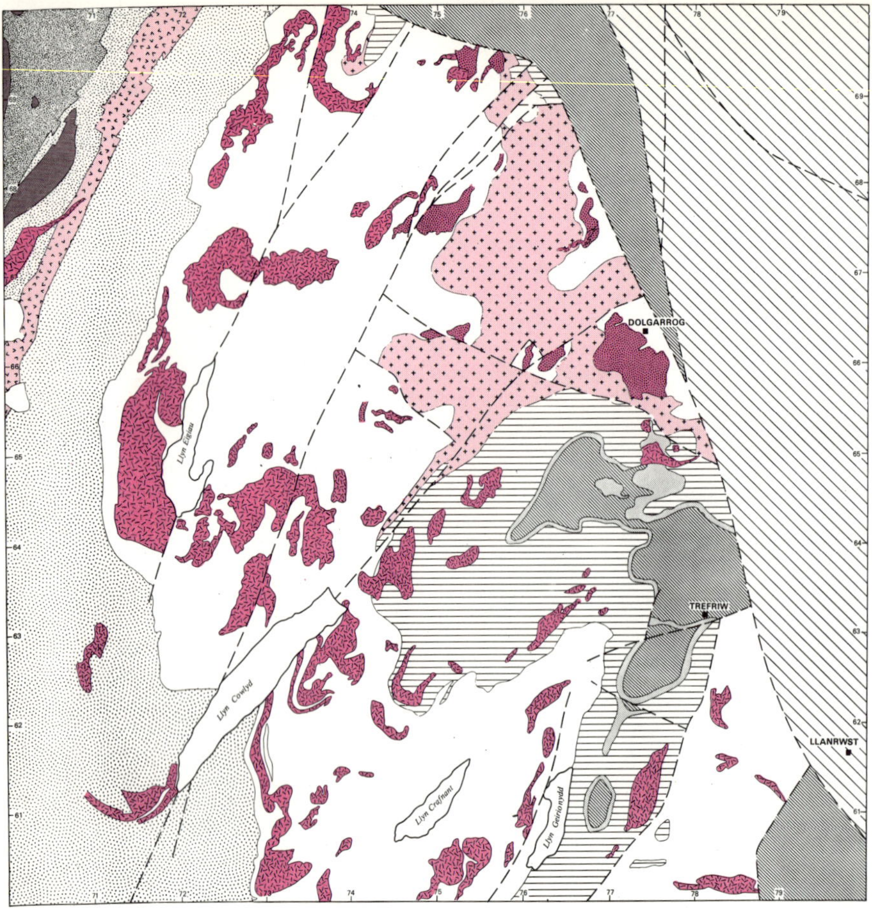

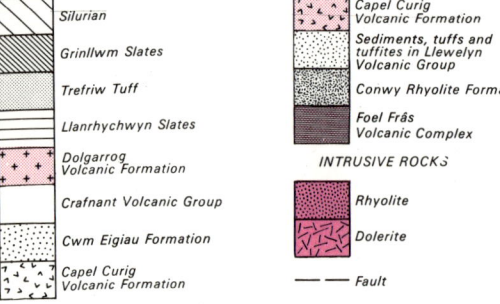

Figure 1
Geological sketch map of the area

Classical areas of British geology

M. F. Howells
B. E. Leveridge
C. D. R. Evans
M. J. C. Nutt

Dolgarrog

Description of 1:25 000 sheet SH 76

INSTITUTE OF GEOLOGICAL SCIENCES
Natural Environment Research Council

London   Her Majesty's Stationery Office   1981

© *Crown copyright 1981*

Bibliographical reference

Howells, M.F., Leveridge, B.E., Evans, C.D.R. and Nutt, M.J.C. 1981 *Dolgarrog: Description of 1:25 000 Geological Sheet SH 76.* Classical areas of British geology, Institute of Geological Sciences. (London: Her Majesty's Stationery Office.)

Notes

National Grid references are given in the form [7000 6880] throughout: all lie within the 100-km square SH.

Numbers preceded by the letter E refer to thin sections in the collections of the Institute of Geological Sciences and numbers preceded by GF, RU, RV, Zl and Zp to specimens in the fossil collections.

Authors

M. F. HOWELLS, B.Sc., PhD
B. E. LEVERIDGE, B.Sc., PhD
C. D. R. EVANS, B.Sc., PhD
M. J. C. NUTT, B.Sc., PhD
Institute of Geological Sciences,
Ring Road Halton, Leeds LS15 8TQ

Her Majesty's Stationery Office

Government Bookshops

49 High Holborn, London WC1V 6HB
13a Castle Street, Edinburgh EH2 3AR
41 The Hayes, Cardiff CF1 1JW
Brazennose Street, Manchester M60 8AS
Southey House, Wine Street, Bristol BS1 2BQ
258 Broad Street, Birmingham B1 2HE
80 Chichester Street, Belfast BT1 4JY

Government publications are also available through booksellers

Institute of Geological Sciences

Exhibition Road, London SW7 2DE

Murchison House, West Mains Road, Edinburgh EH9 3LA

Keyworth, Nottingham NG12 5GG

The full range of Institute publications is displayed and sold at Murchison House and the Institute's Bookshop at the Geological Museum, Exhibition Road, London SW7 2DE

The Institute was formed by the incorporation of the Geological Survey of Great Britain and the Museum of Practical Geology with Overseas Geological Surveys and is a constituent body of the Natural Environment Research Council

Design by HMSO Graphic Design
Filmset by KNP Group, Birmingham
Illustration origination by UDO Litho Ltd.

Printed in England for Her Majesty's Stationery Office by UDO Litho Ltd.

ISBN 0 11 884133 5

Dd696818 K80

Preface

The first geological survey of the district, by J. B. Jukes, W. T. Aveline and A. R. C. Selwyn was on the one-inch scale, published as an Old Series Sheet (78 SE) in 1852. The survey on which the present account is based was carried out by Drs Howells, Evans, Leveridge, Nutt, Warren and Francis between 1968 and 1971 on the six-inch scale, making extensive use of aerial photographs as base maps. The published map (SH 76) covering the Dolgarrog district is the second of a series of sheets on the 1:25 000 scale being produced by the Institute of Geological Sciences to delineate the details of the complex Lower Palaeozoic geology of North Wales. The present account is designed to be read in conjunction with the map.

G. M. Brown
Director

Institute of Geological Sciences
Exhibition Road
London SW7 2DE
1 February 1980

Contents

1 Introduction 1
 Sedimentary rocks 6
 Volcanic rocks 8
 Tectonic setting 8

2 Llewelyn Volcanic Group 10
 Foel Frâs Volcanic Complex 10
 Conwy Rhyolite Formation 12
 Strata between Conwy Rhyolite Formation and Capel Curig Volcanic Formation 15
 Capel Curig Volcanic Formation 17

3 Cwm Eigiau Formation 20

4 Crafnant Volcanic Group 24
 Lower Crafnant Volcanic Formation 24
 Middle and Upper Crafnant Volcanic Formations 33
 Succession south-east of the Crafnant valley 34
 Succession north-west of the Crafnant valley 38

5 Dolgarrog Volcanic Formation 42

6 Llanrhychwyn Slates 47

7 Trefriw Mudstone and Trefriw Tuff 45

8 Grinllwm Slates 51

9 Silurian 53

10 Structure 57
 Folds 57
 Cleavage 59
 Faults 59

11 Mineralisation 61

12 Intrusive Igneous Rocks 65
 Acid 65
 Basic 66

13 Pleistocene and Recent deposits 71
References 76
Excursion itineraries 80
Fossil localities 85
Glossary 87
Index 89

List of illustrations

1 Geological sketch map iv
2 The Conwy valley 2
3 Generalised vertical section 4
4 Generalised ribbon diagram of the Ordovician rocks 7
5 Capel Curig Volcanic Formation section at Llyn Dulyn and Foel Lwyd 16
6 South shore of Llyn Dulyn 17
7 Facies variation in the Cwm Eigiau Formation 21
8 Moraine, Melynllyn 22
9 Correlation of tuffs in Lower Crafnant Volcanic Formation 25
10 Ridge west of Llyn Crafnant 26
11 Photomicrograph, eutaxitic texture, No.1 Tuff, Lower Crafnant Volcanic Formation 28
12 Llyn Eigiau 30
13 Photomicrograph, fine shards, No.2 Tuff, Lower Crafnant Volcanic Formation 31
14 Photomicrograph, coarse to fine shards, No.3 Tuff, Lower Crafnant Volcanic Formation 32
15 Photomicrograph, clast of perlitic-fractured rhyolite, No.3 Tuff, Lower Crafnant Volcanic Formation 32
16 Crafnant valley 35
17 Photomicrograph, vitroclastic fabric, Upper Crafnant Volcanic Formation 36
18 Photomicrograph, pumice clasts and shards, Upper Crafnant Volcanic Formation 37
19 Photomicrograph, vitroclastic fabric, Upper Crafnant Volcanic Formation 38
20 Photomicrograph, hyaloclastite, Dolgarrog Volcanic Formation 44
21 Pillow breccia, Afon Porth-llwyd 44
22 Llanrhychwyn Church 48
23 Grinllwm 51
24 Generalised section Denbigh Grits Group 55
25 Pont Fawr, Llanrwst 56
26 Distribution and orientation of structures 58
27 Lodes and properties in part of the Llanrwst mining field 62
28 Tailings, Parc Mine 64
29 Photomicrograph, perlitic-fractured rhyolite 66
30 Photomicrograph, brecciated perlitic-fractured rhyolite 67
31 Distribution of dolerite intrusions 68
32 Photomicrograph, spilitic dolerite 69
33 Craig Eigiau and Eigiau Dam 70
34 Perched block, Pen y Drum 71
35 Cwm Dulyn 72
36 Valley of Afon Dulyn 73
37 Geophysical traverses, Conwy valley 74
38 Cwm Eigiau 75
Plans of five walks 80–84
Shelly fossils and graptolites 86

Introduction 1

This handbook describes the geology of the district covered by the 1:25 000 map, SH 76. On its eastern side it includes the town of Llanrwst and the villages of Trefriw, Dolgarrog and Tal-y-bont, on the edge of the broad alluvial flat of the River Conwy. Formerly the Conwy broadly divided the counties of Caernarvonshire on the west from Denbighshire on the east, but now the whole of the district forms part of the county of Gwynedd, named after the ancient realm which in the middle of the 13th century extended across most of the high ground of north-west Wales. Much of the area to the west of the Conwy lies within the Snowdonia National Park.

The main physical features are shown in the end-paper at the front of the book. Most of the plain of the River Conwy occurring within the district lies below 7·6 m OD, and the river is tidal up to the vicinity of Trefriw. In the west the ground rises sharply, to approximately 300 m OD above the Conwy valley, and then more gradually to some 900 m OD on the ridge that extends from Drum (771 m OD) to Foel Frâs beyond the western margin of the sheet. Most of this area of high moorland is utilised for sheep and hill cattle grazing. The occurrence of the name 'Hafod' (summer house), for example in the valley of Afon Dulyn, suggests that these upland valleys were areas where transhumance was formerly practised. The Cowlyd valley supported farms and a chapel well into the present century.

The eastern side of the Conwy valley is less precipitous than the west, rising in a series of scarps. Many of the larger of these are wooded while the remaining areas, together with the valley floor, provide the only arable ground in the area (Figure 2).

Forestry was established in the area as early as 1919, when it employed men from the declining mining industry; it is now mainly concentrated in the Nant Gwydir and Crafnant valleys in the south of the district.

The district includes the lakes of Geirionydd, Crafnant, Cowlyd, Eigiau, Melynllyn and Dulyn, all of which have been harnessed at some time as a source of water supply or hydroelectric power. Llyn Cowlyd, the largest of these, is the deepest lake in Wales (69.7 m) (Ferrar, 1973) and is linked by open leat with Ffynnon Llugwy (SH 66) and by tunnel with Llyn Eigiau, providing water for Colwyn Bay and the Dolgarrog Power Station. Drainage from the higher ground is eastwards to the Conwy by Nant Gwydir, Afon Crafnant, Afon Ddu, Afon Porth-llwyd and Afon Dulyn.

Figure 2 Conwy valley viewed from Cefn Cyfarwydd. Trefriw at the junction of the Crafnant and Conwy valleys and Llanrwst on the far side of the alluvial tract

The presence of the Conwy Valley Fault, separating 'Lower Silurian' on the west from 'Upper Silurian' on the east, was first recognised by Sedgwick (1843). Later, the geology of the district was described by Ramsay (1866, 1881) who assigned the rocks on the west side of the Conwy valley to the Bala Series ('Lower Silurian') and those to the east to the Llandovery and Wenlock Series, ('Upper Silurian'). Following Lapworth (1879), the term 'Ordovician System' was gradually introduced to replace the 'Lower Silurian'. In 1889 Harker published the first petrographical description of the igneous rocks.

No further work was published on the district until Sherlock (1919) described the geology and genesis of the pyrite deposit at Cae Coch and Davies (1936) the Ordovician rocks of the Trefriw district, covering much of the area of SH 76. More recently, Diggens and Romano (1968) investigated the succession below the Crafnant Volcanic Group, to the south-west of Llyn Cowlyd, with emphasis on the contained faunas and their zonal significance. Stevenson (1971) described the Ordovician rocks of the Dwygyfylchi–Dolgarrog area, linking the earlier work of Elles (1909) in the Conwy district with that of Davies at Trefriw. The district is bounded on the southern side by the Capel Curig district (Howells and others, 1978) which shows a generally similar Ordovician succession.

The Silurian rocks were described by Boswell and Double (1940) and Boswell (1943) followed by the latter's synthesis of the Wenlock and Ludlow of north Wales (1949). Descriptions of these outcrops are also given by Warren and others *(in press)*.

A simplified geological map is shown in Figure 1, though the reader is referred to the published 1:25 000 map (SH 76), which accompanies this handbook, for details. The generalised sequence, shown graphically in Figure 3, is as follows:

	Thickness
SUPERFICIAL DEPOSITS (Drift)	m
Peat	
Flood-plain alluvium and alluvial cones	
Scree and Head	
Moraine	
Boulder Clay	
SOLID FORMATIONS	
Silurian	
Denbigh Grits Group	750
Pale Slates	?
Ordovician	
Grinllwm Slates	180 to 320
Trefriw Tuff/Mudstone	up to 80
Llanrhychwyn Slates	50 to 400
Dolgarrog Volcanic Formation	50 to 400
Crafnant Volcanic Group	
Middle and Upper Crafnant Volcanic Formations	up to 1000
Lower Crafnant Volcanic Formation	360 to 500
Cwm Eigiau Formation	
Mudstones and siltstones	up to 280
Tuffs and sandstones	up to 365
Mudstones and siltstones	0 to 680
Llewelyn Volcanic Group	
Capel Curig Volcanic Formation	300
Sandstones, siltstones, tuffs and tuffites	350 to 400
Conwy Rhyolite Formation	920
and Foel Frâs Volcanic Complex	100
INTRUSIVE IGNEOUS ROCKS	
Dolerite, trachyandesite, rhyolite	

Most of these divisions are taken from the terminology used by Elles (1909), Davies (1936) and Howells and others (1978). The

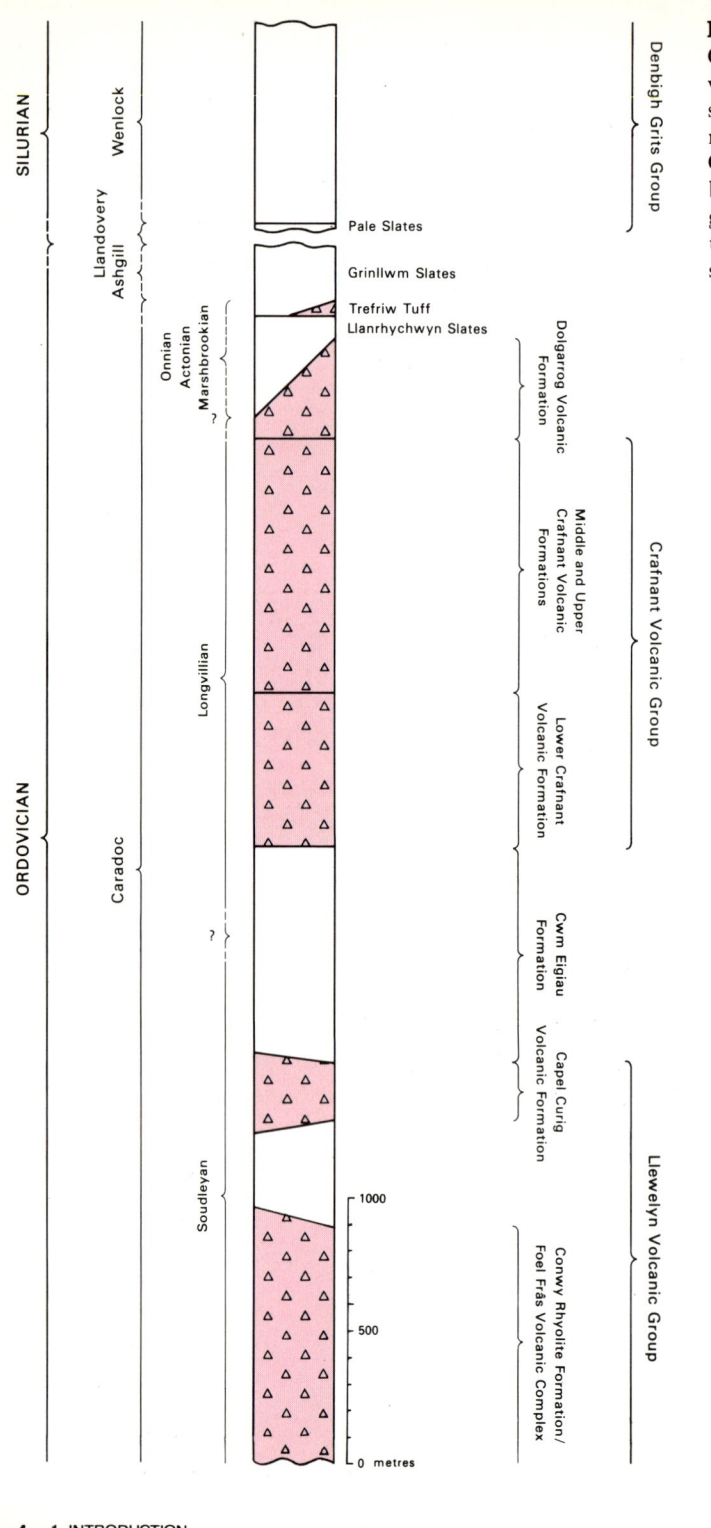

Figure 3 Generalised vertical section showing the relationships in the Ordovician of the broad lithostratigraphical divisions to the Caradoc shelly stages

latter proposed the term Carneddau Group for the whole of the Ordovician below the Snowdon or Crafnant Volcanic Group; however as a result of work to the west and north of the district, it is now proposed that the Carneddau Group should be upgraded to a supergroup and subdivided into the Nant Peris Formation, Llewelyn Volcanic Group and the Cwm Eigiau Formation. The Nant Peris Group does not crop out in the present district. The Llewelyn Volcanic Group here comprises rocks of the Foel Frâs Volcanic Complex, Conwy Rhyolite Formation, and Capel Curig Volcanic Formation (originating from separate volcanic centres). The terms Foel Frâs Volcanic Complex and Conwy Rhyolite Formation are introduced here but are defined by their occurrence in the areas to the west and north respectively (Reedman and others, *in preparation;* Nutt and Leveridge, *in preparation*). The Foel Frâs Volcanic Complex consists of acid and intermediate tuffs, lavas and intrusive rocks, locally thick but of limited lateral extent. The outcrop of the complex led most previous authors—Ramsay (1866), Harker (1889) and Evans (1968)—to the conclusion that it was entirely intrusive. Silvester (1922), however, described these rocks as andesite lavas and tuffs. The Conwy Rhyolite Formation consists of rhyolitic lavas and tuffs, it underlies the Capel Curig Volcanic Formation and the two together form the Conway Volcanic Group of Stevenson (1971) and the Conway Mountain Volcanic 'Series' (less the uppermost 'siliceous ashy grits') of Elles (1909).

The term Cwm Eigiau Formation is proposed here for the mudstones, siltstones, sandstones and tuffs overlying the Capel Curig Volcanic Formation and underlying the Crafnant Volcanic Group. The group is well exposed on Y Lasgallt ridge, between Pen Llithrig y wrâch and Craig yr ysfa on the south side of Cwm Eigiau.

Davies (1936) considered the Crafnant Volcanic 'Series' to be Llandeilian in age and the correlative of the Conway Mountain Volcanic 'Series' of Elles.

It is now apparent that the Crafnant Volcanic Group is higher in the sequence than the Conway Volcanic Group (Stevenson, 1971) and that it thins out to the north of Tal y Fan (SH 77). The term Dolgarrog Volcanic Formation is proposed for the basic volcanics between the Crafnant Volcanic Group and the Llanrhychwyn Slates in the Dolgarrog area (see p. 42).

Fossils occur only sporadically in the rocks of the district and are, in general, poorly preserved, making faunal correlation difficult. Collections support the work of Diggens and Romano (1968), Stevenson (1971) and Howells and others (1978) who concluded that the Soudleyan–Longvillian boundary occurs below the Crafnant Volcanic Group and above the Capel Curig Volcanic Formation. The relationship of the succession to the graptolite zones is less certain though it is apparent that Davies (1936) was incorrect in assigning the Crafnant Volcanic 'Series' to a position between the *Nemagraptus gracilis* and *Amplexograptus arctus* zones (the latter being of doubtful validity in North Wales) as, by comparison with the Shropshire sequence (Dean, 1958), beds no younger than the *Diplograptus multidens* Zone (which is proved to be at least part equivalent to the Harnagian) are present beneath the volcanics.

As in the Capel Curig district, there is uncertainty as to the position of the top of the Longvillian and the presence of the three highest stages of the Caradoc, for the Llanrhychwyn Slates are referred to the *D. multidens* Zone, while the Grinllwm Slates are assigned, at least in part, to the Ashgill on the basis of their correlation with the Bodeidda Mudstones and possibly part of the Deganwy Mudstones (now Conwy Mudstone, Nutt and Leveridge, *in preparation*) at Conwy. The presence of the *Dicranograptus clingani* Zone is uncertain.

In the present district, graptolite faunas in the Silurian have proved the presence of the *Cyrtograptus rigidus* Zone only.

Sedimentary rocks

Reddened conglomerates in the Llewelyn Volcanic Group underlying the Capel Curig Volcanic Formation in the north-west of the district display characters of alluvial fan sheet deposits.

The sediments of the Cwm Eigiau Formation are mudstones, siltstones, sandstones and conglomerates which have yielded sparse brachiopod and trilobite faunas and which for the most part probably reflect a fairly shallow marine environment in which sedimentation broadly kept pace with subsidence (Figure 4).

The sandstones are of greywacke type and for much of the area are laterally impersistent with repeated intercalations of siltstone and sandstone, possibly indicating instability at the margin of a subsiding basin. In the north of the district the Cwm Eigiau Formation consists almost entirely of sandstones, probably fluvio-deltaic and neritic in origin. In the ground to the west of the present sheet, facies variations and current directions indicate derivation from a landmass to the north-west.

Throughout the district there is an increase in the amount of mudstone at the top of the Cwm Eigiau Formation corresponding to the development of a deeper-water environment, as in the Capel Curig district. Within the Lower Crafnant Volcanic Formation the sediments are mainly siltstones and mudstones, the incursions of thick ash-flow deposits not affecting the sedimentary regime sufficiently to cause deposition of coarse clastics. In the central part of the district a richly fossiliferous calcareous band is present between the lowest two ash-flow tuffs at about the level of the local argillaceous limestone noted in the Capel Curig district (Howells and others, 1978).

The lithologies and sedimentary structures in the sediments of the Middle and Upper Crafnant Volcanic formations together

Cyrtograptus rigidus Life size

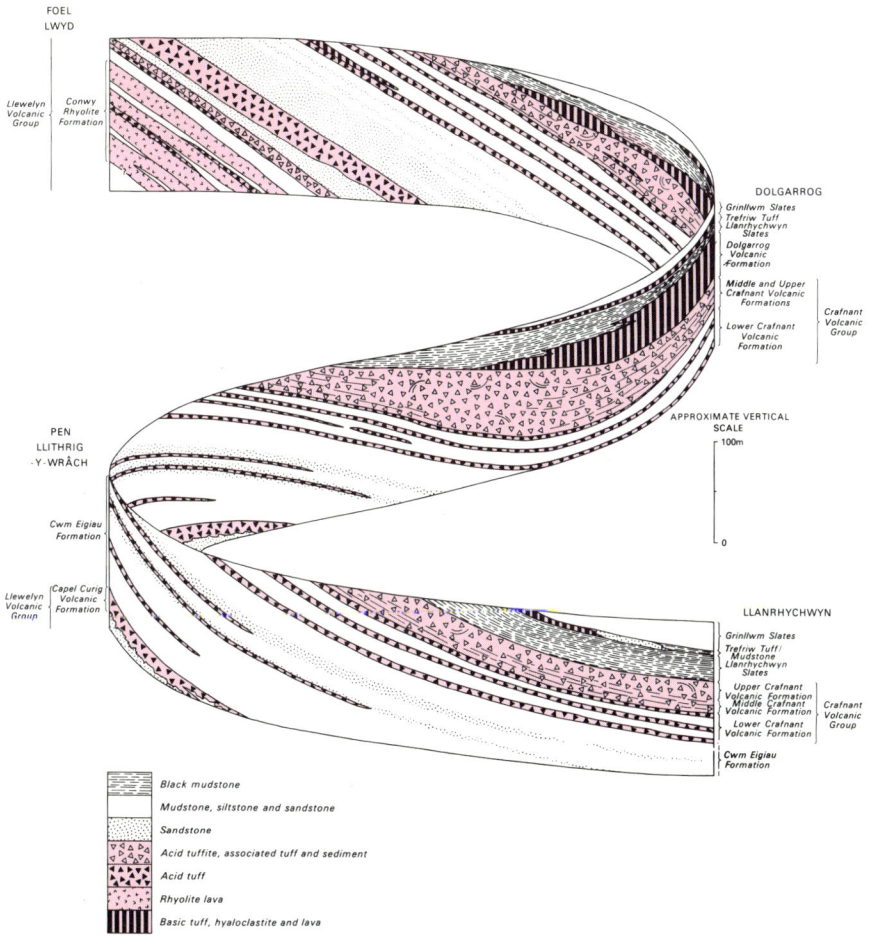

Figure 4 Generalised ribbon diagram showing the lateral variations in the Ordovician across the district. Intrusions omitted.

with the presence of the overlying black slates are evidence of a deep-water environment, the rate of subsidence being greater than that of accumulation. Shallow-water conditions reappeared with the deposition of the Trefriw Tuff which passes laterally into a sandstone and in the south of the district is overlain by siltstones with thin sandstone bands, the Grinllwm Slates.

In the Silurian, the Pale Slates are concealed by drift in the present district. The overlying Denbigh Grits Group comprises greywacke-type sandstones, mudstones and disturbed beds. The mudstones reflect a quiet-water sedimentary environment in which the sandstones were deposited by near-shore slides or turbidity currents and the disturbed beds by down-slope slumping.

Volcanic rocks

Large volumes of both basic and acid volcanic rocks occur within the sedimentary sequence. In the area there are thick accumulations of basalts, basic tuffs and hyaloclastites, though their limited lateral extent supports the assumption that they emanated from local eruptive centres. Their lithologies indicate that the eruptions were mainly submarine, although some volcanoes may have emerged temporarily above sea level. The geochemistry of the basic intrusions and extrusives and their evolution is the subject of recent work by Floyd and others (1976).

The bulk of the volcanic rocks of the district are acid tuffs, consisting of shards and crystals with a minor proportion of lithic clasts and matrix. They exhibit both welded and unwelded textures and are typical products of ash flows (MacDonald, 1972). Associated with them are subordinate amounts of acid air-fall tuffs and tuffites. Many of the acid pyroclastic rocks were formerly described as rhyolites and they have been variously interpreted as submarine and subaerial deposits (Howells and others, 1978). Francis and Howells (1973) postulated that the Capel Curig Volcanic Formation was deposited in a submarine environment in the type area, to the south, but under subaerial conditions in the present district.

The tuffs of the Lower Crafnant Volcanic Formation are mainly unwelded, though composed dominantly of comagmatic material characteristic of ash flows. Their lithology and associated sediment suggests that they are the products of submarine eruptions in which hot suspensive gas was gradually replaced by water to give cool turbid flows. Submarine eruption and emplacement is also invoked for the last major acidic volcanic episode in the district—that forming the Middle and Upper Crafnant Volcanic formations. The lateral variations in thickness and lithology suggest an accumulation of acidic ash flows in a deep-water black mudstone sedimentary environment in the north of the Dolgarrog district with instability in this area causing slumping of the accumulated material on its southern margin. Occasional thin ash flows entered the adjacent deeper-water areas where they were interbedded with undisturbed black mudstones and fine-grained tuffites produced by the subaqueous settling of fine volcanic dust.

Tectonic setting

The district forms part of the Welsh Basin in which thousands of metres of strata accumulated during the Lower Palaeozoic. An Irish Sea landmass (Jones, 1938; Jackson, 1961) along the north-western flank of the basin separated it from a Leinster–Lake District basin. Both basins were ensialic, forming part of a southern Caledonides continental plate which was moving towards a northern continental plate throughout the period, thereby effecting the closure of the ocean named proto-

Atlantic by Wilson (1966) and Iapetus by Harland and Gayer (1972). As further elaborated by Dewey (1969), Powell (1971), Gunn (1973), Jeans (1973) and Phillips and others (1976) the oceanic plate was subducted south-eastwards beneath the southern continent, giving rise to volcanism at various times in both basins. The geochemical model which Fitton and Hughes (1970) erected to relate the volcanism to subduction is perhaps over-simplified. Their suggestion that large volumes of acid material were derived from crustal fusion is not disputed, though as Jeans has indicated, anomalies exist which cannot be explained by this process alone. Whether the acid volcanics were extruded from fissures or from central calderas remains a problem: the only caldera so far postulated in central Snowdonia (Shackleton *in* Beavon, 1963; Rast, 1969; Bromley, 1969) has been disputed (Fitch *in* Bromley, 1969). Certainly there were several eruptive centres. The acidic eruptions were in general of wide lateral extent with marked effect on sedimentation, whereas the basic eruptions were more localised and influenced the sedimentation to a lesser degree.

Extent of the
Llewelyn
Volcanic Group
(see overleaf)

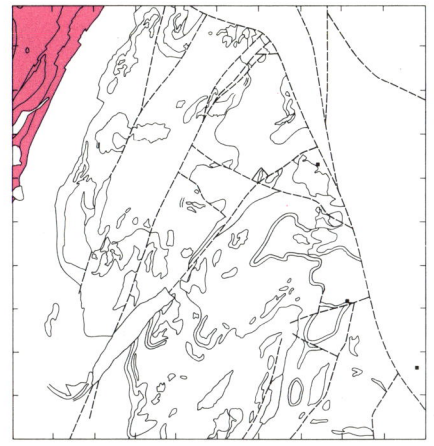
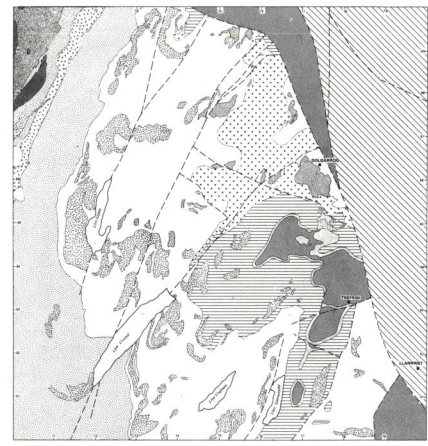

Llewelyn Volcanic Group

2

In the district the Llewelyn Volcanic Group (see p. 4) comprises volcanic rocks emplaced at the edges of two contemporaneous centres which gave rise to the Foel Frâs Volcanic Complex in the west and the Conwy Rhyolite Formation in the north. These are overlain by sandstone and siltstone with subordinate tuffs and the top of the group is defined by the Capel Curig Volcanic Formation.

Foel Frâs Volcanic Complex

Parts of the Foel Frâs Volcanic Complex crop out in the northwest corner of the district. The complex is a volcanic centre association of extrusive and intrusive acid-intermediate rocks. The lavas and tuffs are predominantly trachyandesite in composition (Evans, 1968) and the accompanying porphyrite intrusions are similar in composition and texture to the lavas. Extrusive rocks of the complex are at the same general stratigraphic level as the Conwy Rhyolite Formation although tuffs comparable with those of the Foel Frâs area rest on top of the rhyolites near Bwlch y Ddeufaen, to the north [SH 77]. In the present district the boundary of the extrusive rocks of the complex with the Conwy Rhyolite Formation is a fault, inferred to be penecontemporaneous with the emplacement of the volcanics.

Superficially all rocks of the complex are very similar, being greyish green, fine-grained and porphyritic. Areas of predominantly magmatic or pyroclastic rocks have been delimited to the west, but detailed division is not always feasible in the field.

In the small outcrop of the complex in this district intermediate magmatic rocks and tuffs occur with minor patches of breccia and lenses of acid volcanic rocks. Tuff (E 50276) near the top of the crags south of Afon Anafon [7000 6880], comprises some 25 per cent resorbed euhedral sodic feldspar crystals, up to 2.5 mm in diameter, and 5 per cent mixed lithic clasts in a cryptocrystalline greenish brown matrix. Locally relict shards with quartzo-feldspathic crystallisation show partial welding. In this vicinity [7003 6944] a fine-grained porphyritic magmatic rock (E 46488) contains feldspar phenocrysts, up to 3 mm across and largely altered to sericite and carbonate, and a minor proportion of phenocryst pseudomorphs which are ill defined and consist of aggregates of

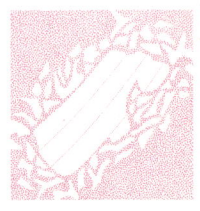

Resorbed sodic plagioclase

chlorite and opaque oxides, suggesting the former presence of primary ferromagnesian minerals. The groundmass is a patchy felsitic mosaic overprinting a cryptocrystalline base in which microlites are flow orientated. The rock is similar both to lavas west of Llyn Anafon and to intrusions in this district.

Adjacent to the boundary fault (see above) there are patches of volcanic breccia and isolated lenses of acid volcanic rocks, predominantly banded rhyolite, though tuff (E 46489) also occurs.

Three discrete bodies of trachyandesite-porphyrite crop out outside the main area of the complex; a small irregular intrusion in the Conwy Rhyolite Formation near the head of Afon Anafon [7032 6892], part of a larger intrusion to the north [7015 6998] and a larger conformable body south-west of Cwm Bychan. The latter is highly altered, its field relations obscured and it is possibly extrusive. However where the alteration is less intense it is petrographically similar to the smallest intrusion.

The porphyrite is generally massive, greyish green, well jointed and locally banded. It forms a line of crags south-west of Cwm Bychan and is well exposed near the northern margin of the district. Feldspar phenocrysts, up to 3.5 mm in diameter, are generally sericitised, chloritised or replaced by carbonate and form 45 per cent of the rock in the smallest intrusion (E 51084) and about 25 per cent in the larger bodies (E 50699, E 50784). In the latter the groundmass varies between hyalopilitic and finely microcrystalline. The original fabric is also discernible through early-stage felsitic recrystallisation in the northern intrusion and also in parts (E 46487) of the largest intrusion but in general coarser quartzo-feldspathic, chlorite and sericite recrystallisation, some of which is a tectonic overprint, has disrupted the magmatic fabric (E 51035) of the latter. However all have opaque oxide minerals evenly disseminated and minor patches of coarser chlorite and opaque oxides dispersed throughout. The link with the lavas of the main mass of Foel Frâs is clearly shown by the textures of the finer rocks (E 50705, E 46488) with dispersed flow-orientated microlites.

Interpretation Recent work (Reedman and others, *in preparation*) shows that thick deposits of lavas and tuffs from the Foel Frâs volcanic centre accumulated to the west of the district in a rapidly subsiding basin partly bounded on the east and north by active faults. Heterogeneity and discontinuity of deposits suggest possible caldera development. Extrusives were largely confined to the volcanic centre although thin representatives of the Foel Frâs Volcanic Complex are present to the south-west and north of the Dolgarrog district. The fault between the Complex extrusives and the Conwy Rhyolite Formation was active during extrusion of the two formations. Bedding attitudes in the Foel Frâs rocks just west of the district indicate primary pretectonic dips into the volcanic centre away from an early fault scarp. Generally a balance was maintained between the two accumulations but occasionally primary or derived deposits may have encroached the main accumulation of the other. Eventually the extrusives overflowed barriers to form the more extensive deposits seen at a similar restricted horizon on SH 66 and above the rhyolites on SH 77.

Conwy Rhyolite Formation

Rhyolites and associated rocks around and to the west of Bwlch y Gwrhyd in the north-west corner of the district form part of the Conwy Rhyolite Formation, (Nutt and Leveridge, *in preparation*). Towards Conwy, where the formation has its thickest development, it comprises two divisions of rhyolite lavas separated by mudstones, tuffite and tuff. In places isolated lenses of black mudstone and acid tuff separate flows but more usually individual flows of massive white to pale yellow banded aphanitic rhyolite are not readily separable. To the west and south of Tal y Fan and into the present area the rocks are sufficiently different in character for them to have been mapped separately on the basis of origin (Davies, 1969) and constitution (Stevenson, 1971). Good exposures in the Bwlch y Ddeufaen area (SH 77) show that this variation in part reflects differentiation within the rhyolite flows, with chlorite and opaque oxide enrichment in green, brecciated crystal-rich tops and bases.

The lavas are referred to as rhyolites (*sensu lato*) due to the overall acidic aspect of their groundmass. Bulk compositions and the degree of sodium metasomatism have not been determined but the present phenocryst compositions suggest that most of these rocks may be termed quartz-keratophyres.

In the Bwlch y Gwrhyd area the formation, up to 920 m thick, comprises rhyolite lavas with intervening mudstone, tuff and tuffite. It is poorly exposed and neither top nor base is seen. Eight flows, each 50 m or more in thickness and some separated by interflow deposits, have been recognised on the south side of Afon Anafon. Faulting and lack of exposure do not permit definite correlation across the area but it is suggested that there are three fairly persistent horizons of interflow deposits and two that are only locally developed.

Rhyolites

The rhyolites are variable in character. The predominant rock type is white to pale grey and forms prominent crags [7020 6885, 7053 6900] on either side of Afon Anafon in the extreme north-west corner of the district. This type commonly grades into a darker greenish grey variety with more abundant phenocrysts. In most cases this occurs in the basal parts of flows [7010 6929] but it may also form separate flows [7041 6938]. Fine flow-banding is ubiquitous and autobrecciation is common especially at flow boundaries where it is associated with vesiculation [7043 6915 and 7047 6903]. In thin section the rhyolites comprise common feldspar and rare quartz phenocrysts in a groundmass of microcrystalline undifferentiated quartz and feldspar, chlorite, white mica and opaque oxides. Colour variation from pale (E47468) to dark greyish green (E 50680) is determined by the content and dispersal of micaceous minerals. The proportion of phenocrysts (up to 35 per cent of the total rock in E 50702) increases with the colour index. The feldspar phenocrysts are albite-oligoclase and occur as individual crystals, up to 2.5 mm, cruciform pairs and small clusters of euhedral and rounded euhedral crystals; carbonate replacement is common. Banding is defined by differing degrees of recrystallisation and proportions of micaceous minerals (E 47483) and its flow origin is suggested by the preferred orientation of scattered feldspar microlites (E 47460). Vesicles are filled by fine quartz mosaic (E 50702) or by peripheral quartz and a chlorite core (E 50680). Perlitic structure is common and suggests that the crystallinity of the groundmass is largely secondary. Recrystallisation fabrics vary, some rocks containing small clear areas of coarse diffuse quartz-feldspar (E 47486) while others show patchy mosaics of larger, up to 0.3 mm, irregularly-polarising composite crystals (E 46541). The recrystallisation is accompanied by segregation of the micaceous component into bands or irregular patches. The extensive sericitisation of rhyolite seen in examples from the cleaved base of a flow (E 47464) and the top of a flow (E 47463) probably reflects a primary alteration process.

Perlitic texture

Interflow deposits

Interflow deposits are mapped at five horizons, the correlation and location of sub-drift boundary lines being partly based on occurrences immediately to the north of the district; volcaniclastic rock overlying rhyolite at the top of the sequence at Bwlch y Gwrhyd is also included within the formation. These horizons are described in ascending order.

The first is represented by a lens of breccia [7013 6983] consisting of mudstone with blocks of banded rhyolite which corresponds with a distinctive flow contact in Afon Anafon [7009 6926].

The second comprises a greyish green cleaved tuff, up to 3 m thick, overlain by dark grey silty mudstone in the Afon Anafon area. The tuff is acidic (E 47466), consisting of rounded and embayed albite-oligoclase and quartz crystals, up to 3 mm, and carbonate pseudomorphs of vesiculated glass fragments, in a micaceous matrix with a strong tectonic fabric. Northwards the mudstone horizon thickens to approximately 30 m [7037 6940] but farther north again it thins and is represented in small outcrops of mudstone associated with acid tuff which cannot be traced through poorly exposed ground.

The third horizon, which is continuous within the district, is some 80 m thick south of Afon Anafon but thins north of the stream [7042 6910], from which point it thickens northwards. South of the Anafon [7002 6868] there is an upward gradation from coarse lithic crystal tuff to finer vitric tuff. The former (E 50274) comprises mixed lithic fragments (50%), including acid tuff, rhyolite, basalt, mudstone, and crystals (10%) of acid plagioclase and quartz in a matrix of fine quartzo-feldspathic shards, chlorite and sericite. The finer tuff (E 50273) consists of plagioclase crystals (5%) dispersed in a mass of mica-replaced shards showing pseudowelding and tectonic deformation. Towards the Anafon the deposit is a heterogeneous mixture of coarse lithic tuff, rhyolite breccia, mudstone and mudstone tuffite. In contrast, greyish green acid tuff forms the entire deposit to the north. Bedding is well-developed with alternating bands and cross-bedded thin flaggy lithic and crystal tuffs. The former (E 47462) has a high clast content (40%), predominantly chloritised perlitic glass, a minor proportion of quartz and plagioclase crystal fragments in a fine matrix with ghost shards. The crystal tuff (E 50703) contains a few clasts and a high proportion of euhedral and rounded albite xenocrysts, up to 1.5 mm across, and fewer quartz euhedra and fragments in a coarser micaceous quartzo-feldspathic matrix with relict shards.

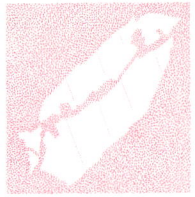

Fractured feldspar crystal

The fourth horizon is variable in thickness and lithology. South of Afon Anafon [7019 6868] some 40 m of massive greyish green crystal-rich tuff separates pale banded rhyolite lavas. It has a clastic appearance and rhyolite blocks are common in its basal part. The tuff (E 46542) contains some 40 per cent albite-oligoclase crystals in a matrix of recrystallised and chloritised perlite fragments and is closely comparable to that occurring at the top of the formation at Bwlch y Gwrhyd. It thins to 5 m north of the stream and here the crystal content is as little as 25 per cent, comprising acid plagioclase crystals, up to 2.5 mm in size, accompanied by a minor proportion of quartz crystal fragments. Chloritised perlitic glass fragments, up to 5 mm, are dispersed in a very fine micaceous matrix. Farther north dark grey silty mudstone, up to 80 m thick on Carnedd Penydorth Goch, overlies an acid tuff (E 50700) [7082 6970] which has a poorly preserved shard fabric showing partial welding and clasts of chloritised glassy rhyolite.

The fifth horizon is represented by grey silty mudstone on the col above Cwm Bychan [7050 6888] and by tuffaceous mudstone [7072 6916] on Bwlch y Gwrhyd. Mudstone occurs at a similar level to the north of the sheet but sub-drift continuity is uncertain.

On Bwlch y Gwrhyd [7077 6911], at the top of the sequence, granulated crystal-rich rhyolite, approximately 40 m thick, is for the most part massive; it rests on pale rhyolite lava. The rock contains 30 per cent euhedral to irregular albite-oligoclase crystals, up to 2.5 mm, (E 47485) in a matrix of recrystallised perlite. The clastic texture is accentuated by the heterogeneous character of the recrystallisation and the variations in the style and deformation of the perlitic structure. The texture is obscured by siliceous recrystallisation and in parts (E 47469) chlorite segregations have produced a pseudo-amygdaloidal form.

Perlitic texture in rhyolite

Interpretation Prior to the emplacement of the Conwy Rhyolite Formation dark grey silty mudstones of the Nant Peris Group (Reedman and others, *in preparation*) were deposited in a marine environment. Similar silty mudstones within the Conwy Rhyolite Formation indicate that it was in this environment that the volcanic rocks were emplaced. The main physical effect on the lavas of contact with water was the quenching of surfaces and the development of volcaniclastic layers composed of glass with perlitic fractures. It is more difficult to determine the chemical effects that this environment produced due to entrapment of water and volatiles and subsequent fumarolic activity, but it seems likely that they caused the compositional variation of the marginal facies. These volatiles probably increased the fluidity of the acid magma making the encroachment of the flows from a centre in the Conwy district more feasible.

Volcaniclastic rocks of the formation include brecciated rhyolite and both massive and bedded tuffs. Local admixture with silty mudstone has produced tuffite. The variability and impersistence of the tuffs suggest the probability of uneven depositional surfaces and local pyroclastic sources and the ubiquity of rhyolite and chloritised perlite clasts within the tuffs indicates emplacement where lavas were being degraded. Bedding in the volcaniclastics may reflect reworking by convection currents and the locally thick deposits of comminuted lava possibly originated by collapse of surface irregularities of lava fronts.

The interflow deposits of the formation are largely restricted to the present district, there being one horizon only to the north of the Aber–Llanbedr Fault (SH 77). This development is evidence of the penecontemporaneous activity of that fault during emplacement of the formation.

Strata between the Conwy Rhyolite Formation and the Capel Curig Volcanic Formation

These strata are about 300 m thick to the west and north of Foel Lŵyd. The lower 130 m is drift covered but a stream section just to the north of the area [7060 7025] exposes grey siltstone overlain by interlaminated siltstone and fine-grained sandstone with separate beds of very coarse sandstone. Higher the sequence becomes tuffitic and at about 75 m from the base a 10-m acid tuff occurs. There is a passage at the top through flaggy-bedded tuffs into banded and flaggy sandy siltstone tuffites and pebbly sandstone tuffites. Above 95 m there is some 35 m of yellow-weathered medium and coarse-grained tuffaceous sandstone.

Coarse crystal lithic tuff (E 50682), up to 40 m thick, succeeding the sandstone is the lowest of these strata exposed in the district. It includes crystals of quartz (8%), albite-oligoclase feldspar (20%) and clasts of siltstone, mudstone and altered lava. The matrix is microcrystalline quartzo-feldspathic with sericite and chlorite in locally varying proportions and a ghost shardic fabric is present throughout. The lack of sorting within the rock, the form of the crystals and the shardic matrix indicates that this is a primary ash flow.

Overlying tuffaceous siltstones (E 50683) and acid tuffites (E 50706) are also poorly exposed but are seen in contact with the tuff north of Foel Lŵyd [7137 6973] and below sandstones in Cwm Bychan [7097 6879]. The siltstone and tuffites are well bedded to the north, though at Cwm Bychan there is a heterogeneous admixture of silty mudstone, shards and crystals.

The sandstones immediately below the Capel Curig Volcanic Formation are conglomeratic and tuffaceous and locally [7064 6781] a fine-grained acid tuff occurs. Bedding is generally ill-defined although massive bedding features are accentuated by erosion on the south side of Afon Ddu. A conglomeratic facies is particularly well developed towards the top of the sequence; it is purple and green in colour, with red-stained pebbles. Thickness of the sandstones varies considerably showing a

Figure 5
Comparative sections of the Capel Curig Volcanic Formation at Llyn Dulyn and Foel Lwyd

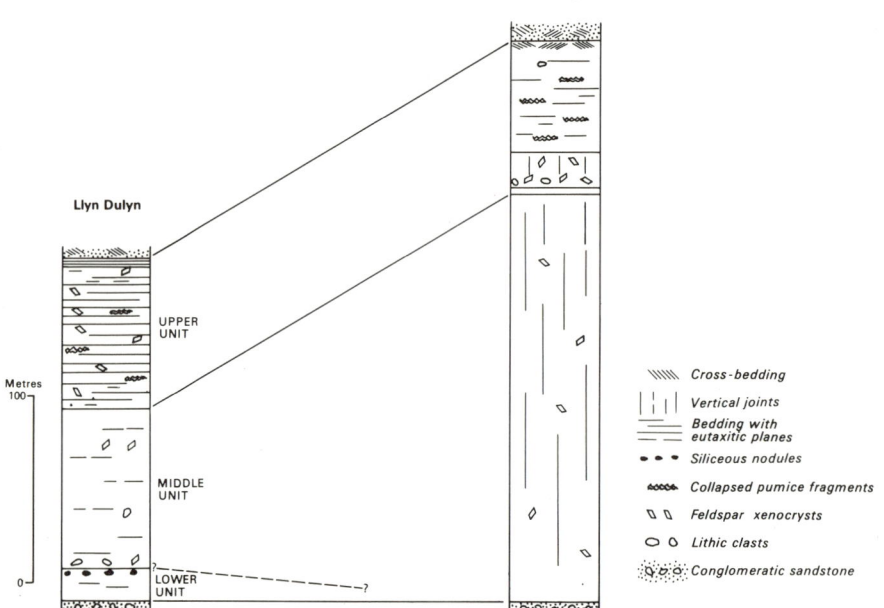

general decrease from about 150 m in the south to some 50 m around Foel Lŵyd.

Capel Curig Volcanic Formation

The formation crops out at Llyn Dulyn and on Foel Lŵyd; between these two areas it is largely obscured by drift.

Llyn Dulyn On the south side of the lake [7002 6632 to 7030 6631] the formation, up to 180 m thick, shows a lower unit with a well-defined zone of siliceous nodules at the top, a central unit with lithic clasts and crystals near the base and distinctive silicification above and an upper unit with well-defined banding (Figures 5, 6).

On the south shore of the lake the lowest tuff, up to 18 m thick, is welded (E 40089) with a good eutaxitic foliation accentuated by concordant segregations of secondary quartz. There is little variation in lithology (E 40090) apart from that in the abundance of tabular feldspar crystals, pseudomorphed by sericite. In the top 5 m, thin-walled siliceous nodules are prominent on the weathered surfaces.

An estimated gap of 5 m in the section coincides with a lithological change at the base of the central unit, up to 87 m thick. The lowest part (E 40091) is rich in lithic clasts which include mudstone, welded tuff, carbonated acidic and chloritised basic intrusives. Crystals of albite-oligoclase and orthoclase are rounded, resorbed and fragmented and commonly altered to sericite and carbonate with conspicuous

Figure 6
Tuffs of the Capel Curig Volcanic Formation, south shore of Llyn Dulyn

included prisms of apatite, up to 1 mm long. Shards are delicate and well defined, set in a sericitic aggregate and deformed around crystals and clasts. Dark green fiamme-like clasts, concordant with the foliation, are particularly prominent on the weathered surfaces and consist (E 40091) of green sericitic mica, chlorite flakes and isolated included altered feldspars, up to 2 mm. These clasts have ragged terminations and their general characters, although obscured by recrystallisation, are those of pumice. Above 34 m from the base of the unit feldspar crystals are less common and lithic clasts are absent. Up to 72 m from the base siliceous crystallisation along eutaxitic planes is particularly prominent, in bands up to 1 cm thick below to 10 cm above. The eutaxitic foliation (E 40094) is partly obscured by the fine recrystallisation of sericite, quartz and feldspar. A well-marked bedding plane defining the top of the central unit can be traced through the outcrop.

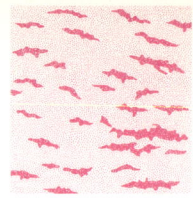

Chloritic fiamme in welded tuff

Eutaxitic planes in welded tuff accentuated by silicification

The upper unit, 75 m thick, is banded with distinctive pumice-rich layers and evidence of upward grading. The recrystallisation fabric (E 40095, E 40096), a quartzo-feldspathic mosaic, obscures the eutaxitic foliation. Isolated sericitised albite-oligoclase crystals occur (E 40097) with lesser amounts of apatite and opaque oxides. Higher in the unit feldspar crystals, up to 3.5 mm are common, as are segregations of quartz, feldspar, sericite and chlorite within the foliation (E 40099). At the top of the tuff (E 40101) is a very fine-grained sericite, quartzo-feldspathic aggregate with a few feldspar crystals and a strong penetrative cleavage fabric.

The upper unit is overlain by 12 m of grey siltstone which in turn is overlain by fine-grained, cross-bedded sandstone.

Foel Lwyd The formation is well exposed on the south side of the hill and a composite section, up to 300 m thick, is shown in Figure 5. The base is not exposed although loose blocks indicate the presence of an underlying conglomeratic sandstone. The formation can be subdivided into a lower massive unit, broadly corresponding to the middle unit of Llyn Dulyn and an upper banded unit directly equivalent to the upper unit of Llyn Dulyn.

The basal 8 to 9 m are not exposed, the lowest exposures consisting of cream coloured, well-cleaved tuff (E 39437) with a tightly packed fabric of small shards replaced by a fine micaceous aggregate. Similar replacement was seen in the basal parts of ash flows of the Lower Crafnant Volcanic Formation in the Capel Curig district (Howells and others, 1973) which suggests that the lower unit of Llyn Dulyn is missing here. The overlying tuff, about 230 m thick, is generally massive with prominent vertical joints and little indication of internal bedding. Recrystallisation is intense (E 39439) with a fine feldspathic aggregate obscuring the original fabric and cleavage accentuated by flakes of green mica. At 180 m above the base of the tuff the cleavage is imposed on a good eutaxitic

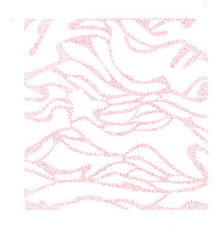

Eutaxitic structure in welded tuff

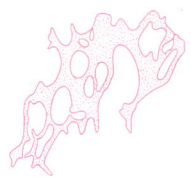

Pumice fragments

fabric (E 39438) with included sericitised albite crystals. At 222 m the eutaxitic fabric (E 39429) is less prominent, a tightly packed mass of variable shards includes complete single bubble and pumice fragments with no indications of collapse; crystals of tabular albite are pseudomorphed by quartz with included apatite. A band, up to 1 m thick, of silty tuffite (E 39426) with small clasts, up to 6 mm, of mudstone and less common altered acid intrusive rock, marks the top of the unit and indicates a break in the emplacement of the tuff.

The overlying tuff, some 57 m thick, has a well-defined basal zone, 3 m thick, of crystal lithic tuff (E 39428). Above, the tuff is massive and well jointed, with isolated pumice fragments and some rounded feldspar crystals, up to a prominent feature, 14.5 m above the base. Above this feature the tuff includes ragged fiamme-like patches of dark chloritic material, possibly collapsed pumice fragments, set in a tightly packed shardic fabric with a eutaxitic foliation, rare crystals of quartz and feldspar pseudomorphed by quartz. At the top the tuff is reworked and overlain by cross-bedded, medium-grained sandstones.

Interpretation   The Capel Curig Volcanic Formation of the Dolgarrog district forms part of a continuous outcrop that extends from near Conwy in the north-east, through the eastern flank of the Carneddau to the ridges of Tryfan, Glyders and Y Garn on the south side of the Ogwen valley (Howells and others, 1978). In the Capel Curig district Francis and Howells (1973) postulated that the tuffs were emplaced in a submarine environment. In the Dolgarrog district the formation is represented by a pile of welded ash flows with no indication of prominent sedimentary partings. On close examination of the sections a subdivision can be made comprising a lower unit of a well-jointed massive ash flow and an upper unit consisting of an accumulation of thin ash flows. The formation overlies coarse mottled green and red conglomeratic sandstone (see p. 16), which certainly represents the shallowest sedimentary environment in the district. On this evidence it is suggested that here the ash flows were subaerially emplaced.

In addition the texture of the tuffs, with silicification accentuating eutaxitic planes and fiamme-like clots, contrasts with the diagnostic characters of submarine emplacement proposed in the Capel Curig district.

The evidence of the mesoscopic and microscopic examination of the two sections suggests that the massive ash flow forming the lower unit was followed by a series of minor ash flows in waning frequency. The reworking of the tuffs at the top of the formation indicates that the pile was eventually submerged and the normal sedimentary environment of the Cwm Eigiau Formation, in which volcanic activity exerted only a minor influence, was re-established.

Cwm Eigiau Formation

3

These strata comprise a variable sequence of sandstones, siltstones and mudstones with subordinate tuffs and tuffites (p. 3 and Figure 7). The thickness varies from about 1400 m in the south to 800 m in the north.

In the south-western part of the district the lowest strata are thin-bedded grey silty mudstones with siltstone bands up to 30 mm thick seen in the stream north of Tal-y-braich-isaf [7020 6030] and on the west side of Cwm Tal-y-braich. A thin band of cream-white bedded vitric tuff is locally included [7025 6225]. At Cwm Tal-y-braich, some 200 m of slates and siltstones separate this tuff from overlying sandstones. On the ridge of Y Lasgallt a thin slumped sandy siltstone is exposed [7055 6260] near the top of the subdivision.

The overlying predominantly sandstone subdivision is best seen at the south-west end of Llyn Cowlyd where it is about 365 m thick. Diggens and Romano (1968) have subdivided these beds into a lower Bwlch Cowlyd Sandstone Formation and an upper Llyn Cowlyd Sandstone Formation, but this practice has not been followed here. Around the south-western end of Llyn Cowlyd the succession is as follows:

f Sandstone, grey banded, current-bedded, locally tuffaceous. Near the middle some 15 m of interbedded agglomeratic tuff contains subangular fragments of acid tuff, siltstone and mudstone in a micaceous matrix and was probably deposited as a mudflow. To the south the sandstones include thin pebbly and conglomeratic bands.   75 to 120 m

e Tuff, fine-grained acid. South of Llyn Cowlyd this tuff is absent though it reappears south-west [724 599] of Craig-wen.   0 to 13 m

d Silty mudstone with thin siltstones, thickens southwards with thin sandstones developed.   up to 175 m

c Sandstone, greenish grey fine-grained and current-bedded; some 12 m near the base contains an abundant shelly fauna (the 'Multiplicata Sandstone' of Diggens and Romano, 1968). This is well exposed on the southern slopes of Pen Llithrig-y-wrâch where it yielded [7167 6144] (loc. 1) a late Soudleyan fauna including *Plaesiomys multifida* [= *multiplicata*]. The latter occurs throughout the sandstones and is not confined to one bed.

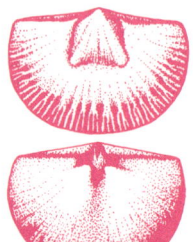

Brachiopod
Plaesiomys multifida
Life size

Figure 7 Facies variation in the Cwm Eigiau Formation

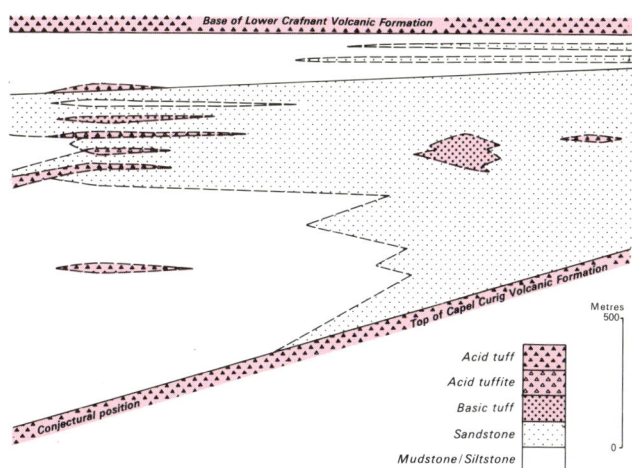

At the top a 11-m fossiliferous tuffaceous sandstone occurs. South of Llyn Cowlyd the unit is thinner and the subdivisions are less easily distinguished.   40 to 90 m

b Tuff, current-bedded, fine-grained acid; locally agglomeratic at the base.   25 to 45 m

a Sandstone, grey or greenish grey massive current-bedded and tuffaceous. This is exposed around Bwlch Cowlyd and at the head of Llyn Cowlyd and in places contains silty mudstone pebbles and some interbedded siltstone and mudstone bands. Southwards, this unit passes into slates and siltstones.   0 to 75 m

This succession may be traced northwards to the north-western side of Pen Llithrig-y-wrâch where two prominent siltstone partings are present within the upper sandstone (f) and a fine-grained acid tuff overlain by siltstones occurs towards the top of sandstone (c). The uppermost acid tuff (e) contains a local development [7106 6241] of agglomeratic tuffs with large carbonate nodules and a rich shelly fauna.

The sandstone continues northwards into Cwm Eigiau and the lowest tuff is seen in small stream exposures [7071 6335]. Farther north on Gledrffordd, isolated exposures include two siltstone and tuff horizons, together with a thin band of mudflow breccia [7057 6455] composed of clasts of mudstone, siltstone and vitric tuff in a muddy siltstone matrix. A broadly comparable sandstone sequence can be traced to the south-west of Melynllyn, although here only the lowest tuff member and one siltstone band can be seen. The bedded and fine-grained tuff has been worked for honestone at the Melynllyn Hone Quarry [7052 6543]. To the east [7093 6542] of the quarry, siltstones yielded a shelly fauna (loc. 2) of Longvillian age (Figure 8).

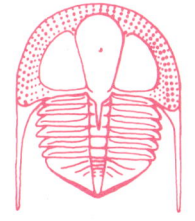

Trilobite *Broeggero-lithus nicholsoni* 1½ times life size

Figure 8 Moraine, Melynllyn. Hone Quarry to the left, on hillside

North-east of Melynllyn the succession is largely obscured by drift although the uppermost sandstones and overlying siltstones are exposed on Clogwyn Maldy. In Afon Garreg-wen [7135 6760] up to 160 m of basic tuffs crop out near the middle of the sandstone sequence. These tuffs are massive to flaggy and range in lithology from a mixture of coarse basaltic lava and pumice clasts, up to 20 cm in diameter, to aggraded basaltic material of siltstone grade. A few basaltic clasts show indented margins, indicating incorporation in the tuff in an unconsolidated state (E 46248). The finer clastic material consists of whole and fragmented albitised feldspars, up to 3.8 mm (E 40106) and some rounded to subrounded quartz in an aggregate of chlorite and feldspar. Near the top the tuff shows a concentration of feldspar crystals, forming up to 60 per cent of the rock, with rare fragments of grey mudstone and softer-weathering pumice. These local basic pyroclastics are comparable in general lithology and stratigraphical position with those of Curig Hill (Howells and others, 1978), though because of drift cover it is impossible to determine if they can similarly be interpreted as forming part of a vent infilling.

On Foel Lŵyd the Capel Curig Volcanic Formation is immediately overlain by sandstones and apart from a narrow drift-covered strip at Cefn Coch, a continuous sequence of some 600 m is exposed.

The beds overlying the sandstone subdivision consist of siltstones and silty mudstones with thin cross-bedded sandstone bands (the Pen Llithrig-y-wrâch Siltstone Formation of Diggins and Romano, 1968), overlain by dark mudstones (the Marian

Cross-bedding

Mawr Mudstones of those authors). These argillaceous beds are about 280 m thick at Gallt Cedryn, but they thin to around 190 m on Llethr Gwyn, south of Llyn Cowlyd. Diggins and Romano (1968, p.39) considered that they were unconformably overlain by the Crafnant Volcanic Group but no supporting evidence for this unconformity was found during the present survey. On Pen y Castell some 176 m of siltstones, with two sandstone bands, lie between the sandstone subdivision and the Crafnant Volcanic Group.

Interpretation The general characters and conditions of deposition of the sedimentary rocks of the Cwm Eigiau Formation have been discussed on p. 6.

Eruption in Snowdonia in Caradoc times

Extent of the Crafnant Volcanic Group

Extent of the Cwm Eigiau Formation

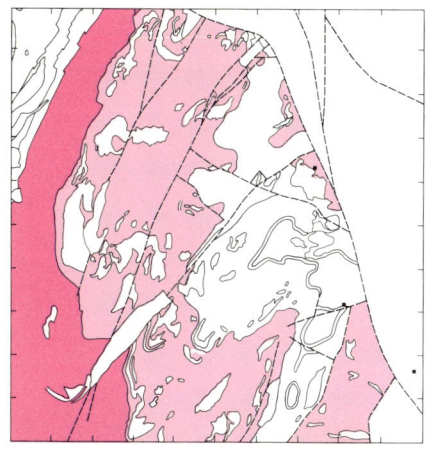

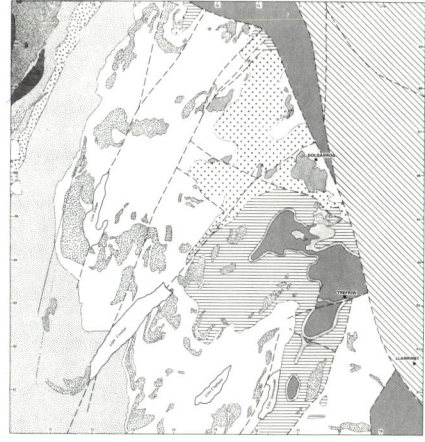

Crafnant Volcanic Group

The Crafnant Volcanic Group is made up of a varied sequence of acid tuffs and tuffites, with subordinate basic tuffs and hyaloclastites and interbedded sediments. The source of some of the acid volcanic rocks was probably in central Snowdonia, to the south-west of the district, but acid tuffs, basic tuffs and hyaloclastites were also erupted from local centres. The group is broadly equivalent to the Crafnant Volcanic 'Series' of Davies (1936) and the component Lower, Middle and Upper formations correspond approximately (Howells and others, 1973, 1978) to the Lower Rhyolitic 'Series', Bedded Pyroclastic 'Series' and Upper Rhyolitic 'Series' of the Snowdon Volcanic 'suite' of central Snowdonia and Dolwyddelan (Williams, 1927; Williams and Bulman, 1931).

The commencement of this volcanic episode can be dated in terms of stages based on the shelly fauna. In the Capel Curig and Betws-y-Coed district the Soudleyan–Longvillian boundary has been drawn within the sandstone-tuffite assemblage some distance below the base of the Crafnant Volcanic Group. In the present area Longvillian faunas have been determined from the Lower Crafnant Volcanic Formation, but little information on the shelly stages has been forthcoming from the higher beds. A middle Caradoc age, in terms of graptolite zones, is established by the *Diplograptus multidens* Zone fauna contained in the Middle and Upper Crafnant Volcanic formations and in the overlying Llanrhychwyn Slates.

Lower Crafnant Volcanic Formation

The Lower Crafnant Volcanic Formation crops out from Llyn Crafnant in the south to the northern boundary of the sheet near Pen y Castell and is in places well exposed. The formation comprises ash-flow tuffs, up to 100 m thick, separated by mudstones and siltstones, calcareous tuffites, hyaloclastites and basic tuffs. In the Capel Curig district (Howells and others, 1973, 1978) three major tuffs, numbered 1 to 3, have been recognised. In the present area (Figure 9) Nos. 1 and 2 tuffs persist, though an intervening tuff (No. 2A) appears south of Llyn Cowlyd and thickens northwards. In the same direction No. 3 Tuff changes character (see p. 32). All the tuffs form prominent scarp features (Figure 10) and are predominantly massive, poorly cleaved grey rocks which are bleached on weathered surfaces. At some localities the basal parts of the tuffs

Figure 9 Sketch map showing correlation of tuffs in the Lower Crafnant Volcanic Formation

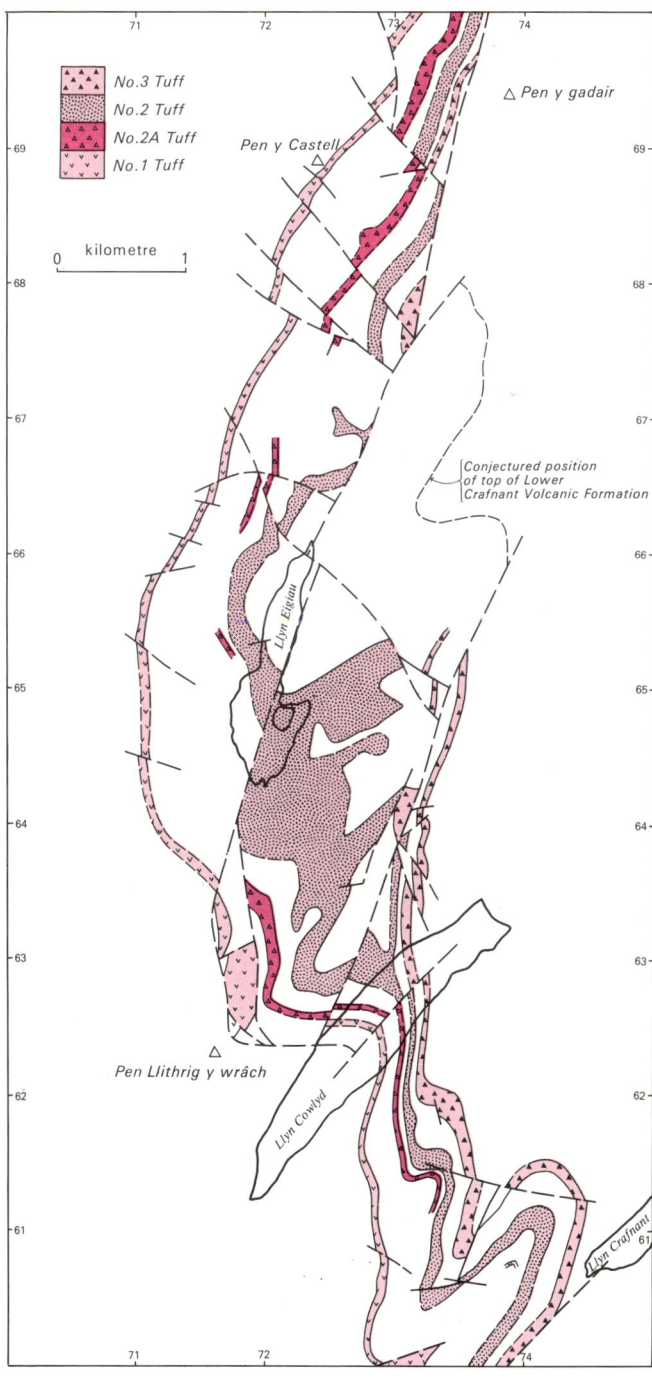

Figure 10 Ridge, west of Llyn Crafnant in tuffs of the Lower Crafnant Volcanic Formation, Craig Wen to Craiglwyn, looking westwards from near Cornel [747 603]

are mica-rich, bluish grey and well cleaved; the tops of Nos. 1, 2A and 2 tuffs are uniformly fine-grained and flinty in character.

The tuffs are all shard-rich and include a variable crystal and lithic content. Most shards fall into the coarse ash range of Fisher (1966), being mainly between $\frac{1}{16}$ and 2 mm and rarely up to 4 mm in diameter and shard shapes are independent of size, varying from broad and massive to delicate and spiky. Welding occurs locally in the No. 1 Tuff in the north of the area. Crystals of fragmented and resorbed albite-oligoclase with subordinate fragmented euhedral quartz are found only in tuffs 1, 2A and 2. Lithic clasts are common in all tuffs and include siltstone, acid tuff, pumice, basic tuff, dolerite and fossil fragments. The matrix is fine-grained quartz, feldspar, sericite and chlorite, probably resulting from the recrystallisation of fine vitric dust. Carbonate concretions are randomly distributed throughout all the tuffs and less common siliceous concretions also occur.

No. 1 TUFF This is some 30 to 50 m thick and prominent throughout the district. In the north it is welded at two localities.

South of Llyn Cowlyd sections up to 30 m thick, in the crags south-west of Craig-wen [7284 6004] and in the cliffs on the south-east side of the lake [7273 6172] were described by Howells and others (1973). Between these sections the tuff is composed of thick regular beds with impersistent siltstone intercalations up to 3 cm thick. Sedimentary clasts are relatively

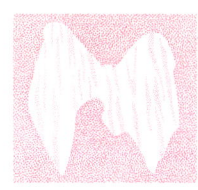

Tubular pumice

Graptolite *Orthograptus calcaratus* Twice life size

Parataxitic structure in welded tuff

few and are concentrated in the lower part. Tubular pumice fragments are prominent on weathered surfaces and are characteristic of the central part of the tuff. Well-defined closely packed shards occur throughout and there is a general upward fining though the range is small. Rounded sub-idiomorphic albite-oligoclase crystals decrease in abundance above the basal layer. The matrix consists of fine-grained greenish white mica, quartzo-feldspathic material and chlorite in varying proportions.

To the north-east of Pen Llithrig-y-wrâch the central and upper parts of the tuff are well exposed. The central part is silicified and fine-grained and the upper part is massive with calcareous bands bearing a rich shelly fauna of Longvillian age [7180 6248] (loc. 3). Farther north, the tuff is well exposed in the cliffs at the eastern end of Gallt Cedryn [7168 6321] where it is 37.5 m thick and generally massive, though with flaggy-bedded parts accentuated by erosion. A crude upward grading is present, with sparse clasts of mudstone, siltstone and possibly basalt restricted to the lowest 3 m. Siliceous nodules are common throughout; though generally small and ill-formed they show a tendency in places to occur in bands. The highest part of the unit consists of bedded silty vitric tuff, with included brachiopod fragments.

North of Cwm Eigiau No. 1 Tuff, about 50 m thick, forms a prominent feature west of the thick dolerite sill of Craig Eigiau and on Cefn Tal-llyn-Eigiau. Here a clast-rich basal zone, up to 22 m thick, massive and silicified in the upper part, is overlain by a thin zone of welded tuff (E 44144). At 1 m from the top the tuff is bedded, with siltstone ribs.

The tuff is well exposed on Pen y Castell where it forms two subdivisions, separated by a few metres of siltstone seen on the north side of the hill. The lower subdivision is about 24 m thick; the lowest 2 m is a well-cleaved greenish grey vitric tuff (E 39435) with delicate finely shaped shards replaced by a fine micaceous aggregate. Siltstone clasts are common, with crystals of altered albite-oligoclase and rounded quartz. This basal zone is overlain by up to 6 m of fine-grained splintery cream-white tuff with siliceous nodules up to 0.3 m long. The tuff (E 39436) is welded (Figure 11), with quartz-feldspar recrystallisation obscuring the outlines of the shards, though a strong eutaxitic foliation, deflected around albite-oligoclase crystals, is discernible. The nodular zone is succeeded by up to 15 m of massive well-jointed fine-grained tuff with isolated albite-oligoclase crystals and chloritic spots. Towards the top the tuff is very fine-grained and recrystallised (E 39433).

The upper subdivision, 9.5 m thick, shows upward grading, with lithic clasts and feldspar crystals more abundant near the base. In thin section (E 39434) the shards of the basal tuff are replaced by mica and show a strong mutual interference; the clast content is similar to that at the base of the lower subdivision

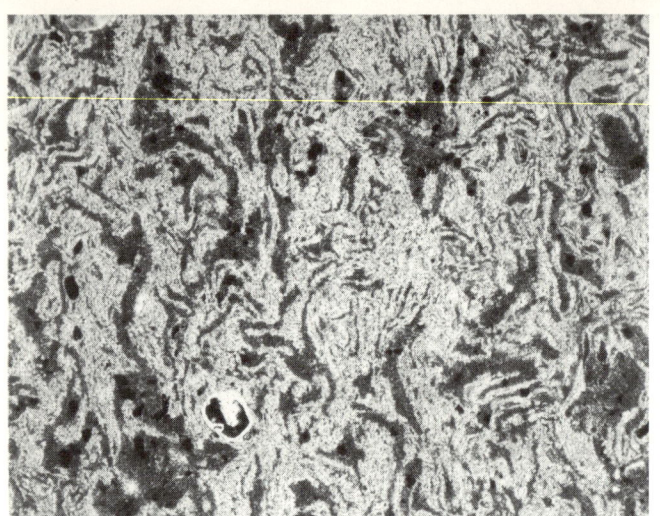

Figure 11
Eutaxitic texture. No. 1 Tuff, Lower Crafnant Volcanic Formation, E 39436. Plane polarised light, × 40

though small quartz fragments also occur. At the top the tuff is reworked and contains fragments of brachiopod shells.

STRATA BETWEEN NO. 1 AND NO. 2 TUFFS   These comprise some 100 to 430 m of siltstone and silty mudstone, with hyaloclastites, basic tuffs and acid tuffs. At Craig-wen [7280 6015] 7 m of cleaved silty mudstone overlie the No. 1 Tuff. Farther north [7285 6104] these sediments are fossiliferous and possibly represent the richly fossiliferous calcareous tuffite at the top of the No. 1 Tuff, north of Llyn Cowlyd (see above).

At the south-west end of the Crafnant valley a lens of hyaloclastite, up to 38 m thick in a stream [7345 6014] near Blaen-y-nant, thins northwards to only 2 m of broken and whole basaltic pillows in a vesiculated and comminuted glassy matrix, within cleaved mudstone. Below Clogwyn yr Eryr [7355 6047] flaggy-bedded basic tuffs up to 5 m thick may represent the lateral reworked material from a local hyaloclastite pile.

On the col between Llyn Cowlyd and Cwm Eigiau the sequence between No. 1 and No. 2 tuffs is faulted. On the north-east side of Gallt Cedryn [7175 6315] the fossiliferous tuffite at the top of No. 1 Tuff (see p. 27) is not recognised, although the overlying sediments are soft-weathered calcareous siltstones with a rich shelly fauna. Silty mudstones are exposed on the north-west side of Llyn Cowlyd and these have been called the Marian Mawr Mudstone Formation by Diggens and Romano (1968), who considered them to lie below the base of the 'Snowdon Volcanic Group'.

North of Cwm Eigiau the strata between Nos. 1 and 2 tuffs are intruded by thick dolerites and exposures in the siltstones are few. On Pen y Castell bedded basic tuffs, agglomeratic in part, overlie No. 1 Tuff; these thin out southwards. Within these strata No. 2A Tuff occurs; it is described separately below. On Pen y gadair greyish green siltstones, below No. 2 Tuff, are exposed in small isolated crags.

No. 2A TUFF   The southernmost exposure of this acid tuff is in the stream section [7388 6074] north of Hendre, where it is bedded. Small clasts and chlorite segregations are common. To the south of Marian Mawr the tuff, up to 6 m thick, forms a dip and scarp feature within siltstones. Here it is bedded, with tuffite bands indicative of an air-fall and water-sorted accumulation. In places [7322 6140] the rock contains weathered-out carbonate nodules, up to 0.5 m, extending into a complex system of narrow carbonate pipes. North of Llyn Cowlyd [7200 6265] the unit consists of 12 to 15 m of cleaved ochreous-weathered tuff below, separated by 2 m of siltstone from an overlying grey silty tuff, up to 6 m thick. A similar division occurs on the south-west margin of the quarry [7190 6348] south of Cwm Eigiau. Farther north the fine-grained acid tuff has been traced east of Cefn-Tal-llyn-Eigiau [7180 6618] and south of Cerrig Cochion [7207 6670].

On Pen y Castell the tuff, up to 40 m thick, is silty at the base (E 44032) with fine shards, replaced by fine white mica, and isolated resorbed albite crystals. Higher (E 44033), the shardic fabric tends to be obscured by recrystallisation. To the north-west of Pen y gadair, the tuff is seen in scattered exposures and maintains its thicker development. It is generally fine-grained and shards decrease upward in size from 0.5 mm near the base (E 39976); they are locally replaced by a chlorite-leucoxene aggregate. Higher in the unit, shardic debris is extremely fine (E 39969) and the texture so even as to indicate an original dust-rich eruption.

No. 2 TUFF   This tuff varies from 35 to 55 m in thickness. It is generally fine-grained and silicified with common siliceous nodules in the upper part.

Exposures of No. 2 Tuff in the lower slopes of the afforested ground west of Llyn Crafnant are confined to small crags and road-cuttings [7410 6085]. Here bluish green, fine-grained vitric tuff shows banding on weathered surfaces, due to the presence (É 40581) of thin impersistent micaceous layers. On the western limb of the anticline north-east of Hendre the tuff forms a strong scarp feature in dense forest and extends south-wards into the crags of Clogwyn yr Eryr. Here, overlying cleaved silty mudstones, the basal tuff (E 40582) consists of prominent shards, a few resorbed feldspars and pumice frag-ments in a dark chloritic matrix.

North of Clogwyn yr Eryr, No. 2 Tuff is generally uniform and massive and up to 36 m thick, with a matrix of fine chlorite and mica in the lower part and a fine quartz mosaic above. The tuff is fine-grained with carbonate concretions randomly distributed throughout and siliceous concretions in general restricted to the silicified flinty top few metres. It is composed of shards of 0.2 to 0.5 mm long, and there is some fining towards the top. Good cuspate shapes are common in the central parts (E 39728) and bubble forms are rare (E 39726). Crystals of albite-oligoclase are more common than quartz, and are in general less than 1 mm diameter.

The tuff forms crags on the south-west side of Clogwyn Du and is here uniformly fine-grained and silicified with prominent vertical joints throughout [7230 6285]. These characters persist on the broad ridge between Llyn Cowlyd and Cwm Eigiau.

North of Llyn Eigiau the outcrop is disrupted by dolerite intrusions (Figure 12). At the northern end of Craig Eigiau the tuff forms a scarp feature adjacent to the dolerite in the main cliffs and is exposed in a faulted outcrop [7260 6660] north of Ty'n-rhos. The tuff maintains its fine-grained character although parts are tuffitic (E 39408) with isolated fine needle-like shards (Figure 13) in a fine chlorite-sericite matrix and ghost-like recrystallised areas probably representing original pumice fragments. Recrystallisation tends to obliterate the original texture (E 39410).

North-west of Pen y gadair No. 2 Tuff, up to 55 m thick, contains a lens of basalt with associated hyaloclastites and is intruded by dolerite. It shows its usual lithological features, including prominent siliceous nodules near the top. In thin section, delicate shards, up to 0.2 mm in length, are replaced by chlorite at the base (E 39965) and grade up into a fine dust-like fraction. In parts (E 39970) there is evidence of pumice clasts in a fine-grained mixture of ragged chlorite flakes and grains of opaque oxides, with isolated irregular patches of coarser recrystallisation. Clearly-defined shard shapes are not always apparent, with the rock tending towards a muddy tuffite

Figure 12 Llyn Eigiau, viewed from near slate quarry [720 636] on the south side of Cwm Eigiau. The ridge beyond the lake consists of dolerite intruding the Lower Crafnant Volcanic Formation

(E 39971) or siltstone (E 39980) with a strong cleavage fabric.

STRATA BETWEEN NO. 2 AND NO. 3 TUFFS These beds, some 30 to 75 m thick, are mainly mudstones, though thin basaltic tuffs occur locally. In the stream section [7383 6108] east of Graig-lwyd they include 13 m of basaltic tuff (E 40573) composed of a fine-grained mass of chloritised glass with carbonate patches, ilmenite and albitised idiomorphic feldspar crystals, up to 2 mm long, overlying black cleaved silty mudstones. The tuff is well bedded at the top (E 40574) with coarser and finer bands of basaltic pumice fragments; it can be traced around the axis of the syncline, to the south-west of Graig-lwyd, to the hill east of Craiglwyn where it is well-bedded throughout. To the north, the tuff is exposed south of Creigiau Gleision [7339 6165]. On Creigiau Gleision these beds consist of dark grey cleaved silty mudstones. On the north side of Llyn Cowlyd a band of bedded basic tuff, up to 2.5 m thick, crops out locally [7308 6360] east of Craig Ffynnon.

NO. 3 TUFF This tuff, up to about 100 m thick, is heterogeneous with a wide range in shard shape and size; it is commonly rich in clasts and contains tuffitic and silty intercalations.

Around Llyn Crafnant the tuff is well exposed, coarse-grained, shard-rich and agglomeratic in parts (E 40575, E 40579) with acid pumiceous and basic lava clasts. Shards vary in shape from fine elongate and cuspate to squat bubble walls and in size from less than 0.1 mm to 4.2 mm. At Graig-lwyd the tuff shows a wide range of shard size (E 40582) (Figure 14), with parts showing a strong mutual interference and distinctive clasts (E 40585) of perlitic rhyolite or welded tuff (Figure 15).

To the north, the tuff forms a prominent scarp feature on Creigiau Gleision where it reaches its maximum thickness. The

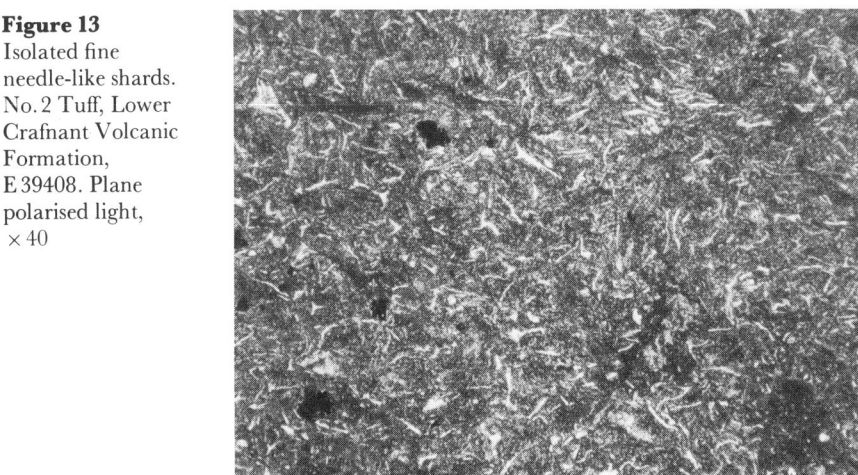

Figure 13
Isolated fine needle-like shards. No. 2 Tuff, Lower Crafnant Volcanic Formation, E 39408. Plane polarised light, × 40

Figure 14
Vitroclastic fabric, coarse to fine shards, No. 3 Tuff, Lower Crafnant Volcanic Formation, E 40582. Plane polarised light, × 40

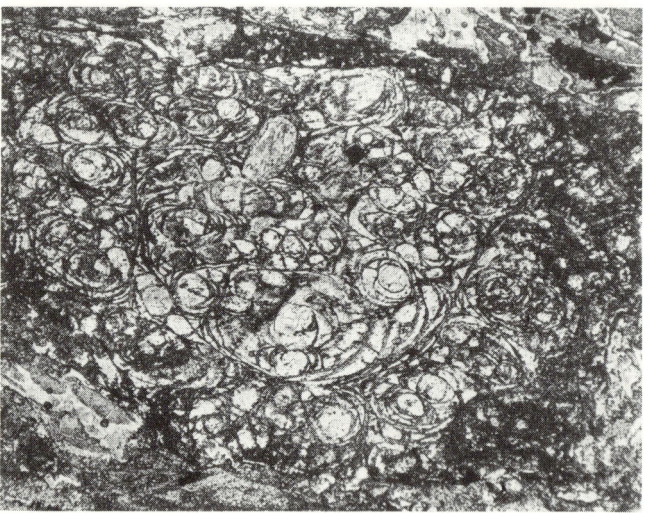

Figure 15
Clast of perlitic-fractured rhyolite or welded tuff. No. 3 Tuff, Lower Crafnant Volcanic Formation, E 40585. Plane polarised light, × 40

base is here gently undulating and the basal lithic tuff contains clasts of pumice, vitric tuff and less abundant basic tuff and siltstone. The main part of the tuff is similar in lithology and texture to that already described.

To the north of Llyn Cowlyd the tuff, though faulted, can be traced east of Eilio [7344 6481]. Exposures in a line of small crags to the west of Pen y gadair are thought to be No. 3 Tuff on the grounds of their relationship with the well-defined underlying No. 2 Tuff, though their lithology is markedly different. The tuff here is recrystallised and contains rounded and embayed albite-oligoclase crystals (E 39978) and in places (E 39977) the rock is muddy and contains widely dispersed recognisable shard fragments. Lithic clasts include chloritised basic intrusives, fine siltstones and perlitic rhyolite or welded tuff.

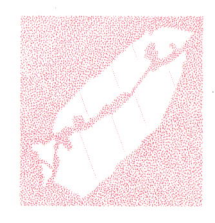

Fractured feldspar crystal

Interpretation The tuffs of the Lower Crafnant Volcanic Formation of the present district are typical products of ash-flow eruptions. Their association with marine sediments and included marine fossils are clear evidence that they were emplaced in a submarine environment. The occurrence of layers of pumice, blocks and siltstone intercalations within the tuffs are primary flow features produced by the loss of energy within the ash flows and are characteristic of their distal portion (Smith, 1960). In much of the area the tuffs are unwelded which, considered together with the lack of induration of the included clasts and the preservation of fossils, suggests that the flows lost most of their initial heat before final deposition. The presence of an undoubted welded fabric at Pen y Castell (but nowhere to the south) indicates that the flow was here insulated, probably by a carapace of quenched material. This evidence would support the postulation by Howells and others (1973) that these flows are the direct products of eruption and not secondary accumulations from the sloughing of unstable masses of pyroclastic debris on the flanks of a submarine volcano, as described by Fiske and Matsuda (1964). The sediments associated with welded tuff at Pen y Castell are identical to those associated with No. 1 Tuff throughout the district and there is no evidence to support a subaerial emplacement at this locality.

The ash flows of the Lower Crafnant Volcanic Formation are widespread in north-eastern Snowdonia and are the lateral equivalents of the Lower Rhyolitic 'Series' (or Lower Rhyolite-Tuffs) of the Snowdon and Dolwyddelan sequences (Williams, 1927; Williams and Bulman, 1931; Beavon, 1963). The details of this correlation (Howells and others, 1973) include evidence to show that Nos. 1 and 2 tuffs were derived from the west and No. 3 Tuff from the north or north-west. In the Dolgarrog district No. 2A Tuff increases from a thin banded fine-grained flinty tuffite south of Llyn Cowlyd, to a typically shard-rich ash-flow tuff, up to 40 m thick, north of Craig Cefn coch. This indicates a source to the north or north-west similar to that postulated for No. 3 Tuff.

The presence of local eruptive centres is indicated by thick piles of basic volcanics between, and in marked contrast to, Nos. 1 and 2 tuffs on the ridge south-west of Llyn Crafnant and at Pen y Castell. These hyaloclastites and pillow basalts are characteristic of a subaqueous environment and, unlike the adjacent acidic ash flows, are of a restricted lateral extent.

Middle and Upper Crafnant Volcanic formations

In the Capel Curig district the Middle Crafnant Volcanic Formation (Howells and others, 1971, 1973) comprises a flaggy to massive tuff, tuffite and mudstone sequence with a few thin turbiditic sandstone bands. The Upper formation (equivalent to the Upper Tuff Bed of Davies, 1936) is made up of a massive ill-sorted mixture of pyroclastic debris and mud with no well-defined bedding. In the present district this subdivision is only possible in and to the south-east of the Crafnant valley where the Middle Crafnant Volcanic Formation is about 230 m thick and the Upper some 40 to 100 m; farther north the two formations cannot be separated.

South-east of the Crafnant valley

Middle Crafnant Volcanic Formation

In the Nant Gwydir valley, south-west of Llanwrst, exposures in the formation are few and are obscured by dense afforestation. In the faulted blocks around Whitebarn and at Coed Clogwyn-gwlyb thick tuff/tuffite horizons form prominent features in a dominantly slate-siltstone sequence. The overall lithologies are similar to those of the type area around Sarnau (Howells and others, 1978). Near Llety [7850 6058] a coarse ash-flow tuff contains rounded clasts, up to 0.3 m, of dark grey vitric tuff set in a grey argillaceous matrix with common rounded albite-oligoclase and isolated euhedral quartz crystals (E 40564). Pumice clasts are generally common (E 40563, E 40564) and carbonate nodules (E 40565) randomly distributed.

Fine-grained vitric tuff with isolated tuff and tuffite blocks can be traced north-west of Whitebarn and interbedded tuffs and black cleaved mudstone are exposed in a stream east of Bwlch [7844 6160]. The association of these tuffs with bedded silicified tuffite and splintery mudstone compares closely with the Sarnau sequence.

Between Llyn Geirionydd and Afon Crafnant fine-grained vitric and dust tuffs make up much of the Middle Crafnant Volcanic Formation, forming thick flaggy and massive beds. Near the northern end of Llyn Geirionydd thick flaggy tuffs and tuffites are well exposed and at one locality [7622 6156] contain large carbonate nodules, up to 1.5 m in diameter, commonly joined along the bedding planes. The nodules weather out leaving large cavities. In the scarp west of Glan Geirionydd Mine [7605 6130], slates, vitric tuffs and tuffites are exposed. Towards the top of this scarp there is an upward passage into grey slates with tuffaceous bands. These are overlain by a persistent band of cleaved black mudstone, up to 20 m thick, which underlies the muddy tuff of the Upper Crafnant Volcanic Formation. The slates have been worked in Clogwyn-y-Fuwch Quarry [7600 6168], where the base of the Upper Crafnant tuff forms the hanging wall of the excavation (Figure 16).

In a tributary [7640 6260] on the north side of Afon Crafnant the upper part of the formation is composed of muddy tuffs with feldspar crystals and banded tuffites interbedded with cleaved mudstones. To the south-west, this part of the sequence forms the core of the anticline west of Cae-crwn. A typical evenly-bedded section of banded tuff and tuffite with silty mudstone intercalations, is seen in crags farther to the south-west

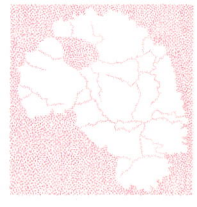

Quartz xenocryst

Figure 16
View south-westwards from near Pen yr Allt. Mynydd Deulyn left of centre, with dumps of waste from excavation in slate underlying the Upper Crafnant Volcanic Formation. To the right, the Crafnant valley, Allt Goch and Clogwyn yr Eryr, the ridge in the distance

[7514 6180]. Banded mudstones with crystal tuffs and tuffites crop out on the east bank of Llyn Crafnant [7517 6118]. Near Cornel [7487 6057] beds of crystal-rich tuff, up to 4 m thick and muddy in places, occur within cleaved mudstones. Higher in the sequence, in the nearby quarry [7465 6030] black pyritic cleaved mudstone overlies tuffaceous mudstones and muddy tuffs with spheroidal carbonate concretions up to 0.5 m in diameter. A quarry [7428 6013] south of Coed Maes-mawr shows some 12 m of black pyritic slates.

On the north-western side of Llyn Crafnant, below Allt-goch [7485 6165], up to 15 m of evenly bedded laminated tuffites and mudstones are exposed and the beds at the base of the formation are overturned in a forestry road-cutting [7447 6107]. Here the tuff is overlain by black cleaved mudstones, up to 9 m thick, and these in turn by 15 m of tuffaceous mudstone with bands of feldspar crystals and mudstone clasts.

Upper Crafnant Volcanic Formation
South-west of Llanrwst the Upper Crafnant Volcanic Formation forms the scarp of Clogwyn Gwlyb and crops out extensively in the afforested area immediately to the north. Between Llyn Geirionydd and the Crafnant valley the formation can be traced around an anticlinal structure, but to the north-west of Allt-goch its base becomes progressively less well defined.

In the faulted area between the Llyn y Parc Fault in the east and the Pen y ffridd Fault in the west the formation consists predominantly of grey muddy crystal tuffite, well cleaved and poorly bedded or unbedded. A matrix of sericite and chlorite is prominent throughout and commonly gives rise to an associated strong cleavage fabric (E 37175). Shards, crystals and pumice

Figure 17
Vitroclastic fabric, shards in iron ore-enriched matrix. Upper Crafnant Volcanic Formation, E 37184. Plane polarised light, × 40

fragments are variously distributed within the matrix and show no signs of sorting. The base is commonly covered by scree though it is seen overlying black slates in the narrow gorge at Coed Ffrith-Sian [7895 6104].

Near the base of the scarp at Clogwyn Gwlyb [7834 6060] the tuffs contain decomposed carbonate nodules, up to 1 m long, randomly distributed apart from a concentration near the base. Here the rock (E 37188) consists of ill-defined coarse to fine shards, devitrified and in places replaced by a fine chloritic aggregate. Above the crags is a vitric tuff (E 37184) with rare rounded albite-oligoclase crystals and clasts of dolerite, welded tuff, pumice and mudstone. The matrix is almost entirely iron ore (Figure 17) and is possibly of hydrothermal origin, the locality being close to Rhuad y Rhwst Mine. A similar though more patchy development of iron ore in the matrix has been seen at about the same horizon near the Hafna Mine [7802 6043] and at Coed Cefn-maenllwyd [7840 6134].

Pumice clasts, some tubular, up to 8 cm long and randomly distributed, are prominent in places on weathered surfaces, as at Bwlch-y-gwynt [7767 6034] where iron mineralisation has been accentuated by weathering. The tuff (E 38923) is shard-rich with tightly packed areas showing slight welding (Figure 18). North of Pen Ucha'r-gwaith small exposures in the tuffite/tuff show the heterogeneous character of the lithology, the variation extending from tuffaceous mudstone with isolated feldspar crystals and cuspate shards to a tightly packed vitroclastic fabric with pumice clasts. On the western slopes of Pen Ucha'r-gwaith exposures near the top of the formation consist of tuffaceous grey slate with scattered decomposed feldspar crystals. Cleaved grey silty tuff at the roadside [7740 6053] includes large clasts, up to 1.5 m in diameter, of pumice with a distinctive ferruginous weathering. Farther north, on the south side of Pant-y-carw

Figure 18
Pumice clasts and shards, slightly welded. Upper Crafnant Volcanic Formation, E 38923. Plane polarised light, × 40

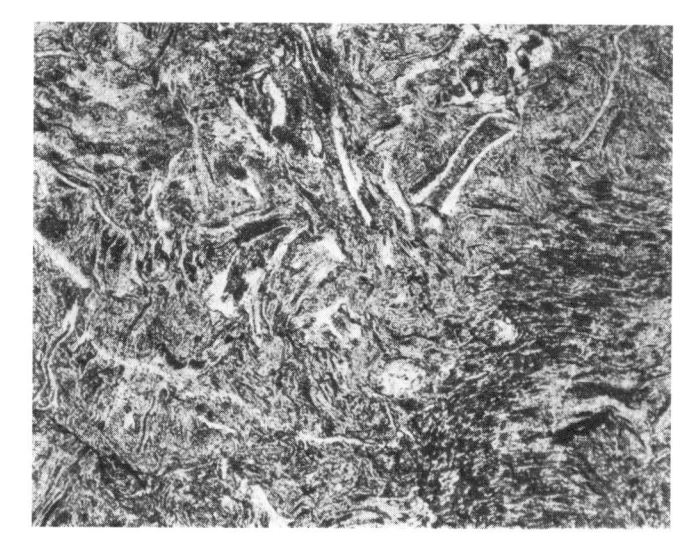

Quarry [7829 6180], silicified tuffs contain isolated fragments of mudstone, up to 0.3 m, and tubular pumice, up to 6 cm.

Muddy bluish grey crystal tuffs with mudstone clasts, up to 0.5 m diameter, and small carbonate concretions are exposed [7625 6112] east of Llyn Geirionydd Mine where they overlie black cleaved mudstones of the Middle Crafnant Volcanic Formation. West of Afon Geirionydd the tuffs are coarse in places and contain shards up to 3.5 mm (E 39755), rounded and resorbed albite-oligoclase crystals up to 2.2 mm, and clasts of pumice, basic tuff, and chloritised siltstone (E 39754) up to 5 mm. In Afon Crafnant, near Gellilydan [7689 6288], and on the edge of Gelli Plantation a compact vitric tuff is exposed. On the north side of the Crafnant valley [7590 6245] cleaved muddy crystal tuffite with thin bands of flaggy-bedded silty tuffite crops out.

At the north end of Llyn Crafnant [7535 6170] an ill-defined flow unit, 26 m thick, has been recognised within the tuff sequence. Clasts of dolerite, spilitic basalt and chloritised siltstone are concentrated near the base (E 39690, E 39691), with pumice and welded tuff more prominent higher in the unit. The matrix, varying between muddy and siliceous, increases towards the top of the flow to give a crude upward grading. This unit is overlain by a typically heterogeneous muddy crystal tuff (E 39605), which contains blocks of mudstone up to 0.6 m and lies near the top of the formation.

The formation crops out in crags on the north side of Mynydd Deulyn and the base is exposed in quarries and caverns in the underlying slates. In the synclinal outlier near Cynllwyd-bâch, small exposures of muddy crystal tuff are common on the eastern limb but elsewhere the tuffs are clean and vitric (E 39599) (Figure 19) with shards ranging from 0.1 to 3.5 mm in length and rare albite-oligoclase crystals. In the forestry

Figure 19 Vitroclastic fabric. Upper Crafnant Volcanic Formation, E 39599. Plane polarised light, × 40

road [7533 6096], south-east of Cynllwyd-bâch, a section near the base of the formation shows a thick tuffite with an irregular base and a massive lower zone grading upwards into silty tuff and black mudstone.

North-west of the Crafnant valley

Undivided Middle and Upper Crafnant Volcanic formations

Davies (1936, p. 73) in describing the Upper Tuff Bed (Upper Crafnant Volcanic Formation of the present account) recognised that to the north-west of Allt-goch 'the argillaceous content diminishes rapidly and it passes into a rhyolitic crystal and lapillar tuff'. With this change it is no longer possible to separate Middle and Upper Crafnant Volcanic formations and the two are considered together.

South-east of Creigiau Gleision, mudstones and siltstones at the base of the subdivision are commonly tuffaceous and contain patches of feldspar crystals. In the scarp at Creigiau Gleision 7 m of siltstones overlie No. 3 Tuff and underlie a dolerite sill. Above the sill are 30 m of tuffaceous slates, tuffites, banded grey vitric tuffs, coarse-grained in places, and 6 m of flaggy-bedded grey basic tuff capped by an upper dolerite sill. This sequence is closely comparable with that of the Middle Crafnant Volcanic Formation in the ground to the south. At the north end of Creigiau Gleision the fine-grained tuffs and tuffites are indurated, which is partly the result of silicification but also the effect of contact alteration by dolerite intrusion.

South-west of Pen y Craig-gron clean white-weathered silicified tuffs with clots of crystals are common at the base of the sequence; they contain partings of impersistent cleaved

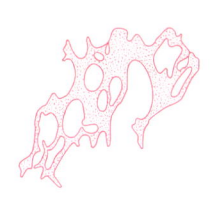

Pumice fragment

siltstones and tuffaceous mudstones. Higher beds are more argillaceous and there is an upward passage into crystal lithic tuffs. Fine-grained silicified vitric tuffs, tuffites and tuffaceous mudstones occur in approximately equal quantities. Though most of the varied lithologies are well defined, bedding is poor.

On the south-east side of the Llyn Cowlyd dam the uppermost part of the formation includes vitric tuffs, some agglomeratic pumice-rich tuff and, near the top [7436 6309], banded tuff and tuffite with layers of feldspar crystals.

To the east of Craig Ffynnon the lowest part of the formation comprises 30 m of massive basic tuffs, thin-bedded at the top. Northwards [7335 6438] the tuffs thin and are underlain by a few metres of mudstone. Here the basic tuff (E 39673) includes fragments of highly-altered basaltic pumice. The lowest vitric tuff, fine-grained and silicified (E 39672), forms a persistent horizon near Craig Ffynnon, overlain in the south [7345 6365] by bedded tuffs and tuffites and a few coarse tuff bands with scattered blocks. On the east of Eilio these beds are less well exposed, though a thin band of basic tuff occurs [7360 6465] near the base. To the north of Siglen up to 260 m of acid tuff, tuffite and mudstone shows confused bedding and the admixtures of lithologies exist down to a microscopic scale. As the wedges of cleaved mudstone cannot be traced laterally it is inferred that mudstone deposition was disturbed by repeated incursions of tuff into a deep-water environment or by post-depositional slumping, or both.

On Moel Eilio tuff/tuffite horizons in places [7450 6548] form crude scarp features with depressions representing the black mudstone intercalations. Lithologies range from acid vitric tuffs through disturbed beds of tuff and mudstone to complete admixtures of tuffs with mudstone clasts, which in many instances are orientated along the cleavage planes. Blebs of basalt (E 39417) in the tuffs are altered to chlorite, ilmenite and sericite and irregular vesicles filled with quartz and chlorite are the only remnant of the original fabric. Two impersistent bands of mudstone, locally highly contorted and in places tuffaceous, can be traced on the flanks of Moel Eilio.

North of Afon Porth-llwyd the Middle and Upper Crafnant Volcanic formations, about 1000 m thick, are much obscured by drift. Beds probably near the base are intermittently exposed in Afon Dulyn where black pyritous mudstones [7380 6763], with a graptolite fauna of the *D. multidens* Zone, are seen. Above, the mudstones, some 120 m thick, are interbedded with tuffites up to 1 m thick, varying from ill-defined disturbed beds of pyroclasts with mudstone bands to well-defined beds of totally mixed pyroclasts and mudstone which are most commonly fine-grained and flinty. Some of the finer beds are laminated pale and dark grey and resemble lithologies of the Middle Crafnant Volcanic Formation of the Sarnau area (Howells and others, 1978). Within this sequence [7394 6778] is a bed of basic tuff of siltstone grade, but with coarser pumiceous layers. Bands within the tuff are cross-laminated, with deformed loaded ripple marks. In thin section (E 39943) small cuspate fragments, less than 0.1 mm in length, indicate derivation by comminution of hyaloclastite.

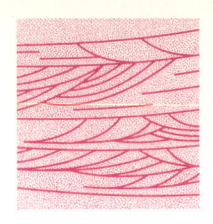

Outcrops of succeeding rocks in Afon Dulyn are disconnected and intruded by extensive dolerite sills. The lower of two fine-grained acid tuffs [7412 6791, 7467 6816], contains a high proportion of feldspar crystals and lithic clasts. Muddy tuffite (E 39944) overlying the lower tuff comprises pyroclastic and mudstone debris with pumice fragments, feldspar crystals and lithic clasts occurring in clusters throughout the rock. Basic, coarse and fine-grained thin-bedded tuffs overlie the upper of the acid tuffs and are in turn succeeded by laminar-bedded acid tuffs and tuffites. Exposures in crags approximately 200 m to the south [7474 6808], near Rowlyn-uchaf, comprise pillow lavas, hyaloclastites and basic tuffs which pass laterally into acid tuffs, black mudstones and locally tuffaceous grey siltstones. The acid tuffs, up to 20 m thick and laterally impersistent, are fine-grained with much recrystallisation (E 41452). Above, acid tuffs including a major unit up to 65 m thick are overlain by mudstones, exposed in small crags [7488 6790] east of Rowlyn-uchaf. The tuffs show an upward grading in shard size from 0.5 mm near the base (E 41442) to 0.2 mm near the top (E 41454). Clasts are predominantly of pumice with a few recrystallised rhyolitic fragments.

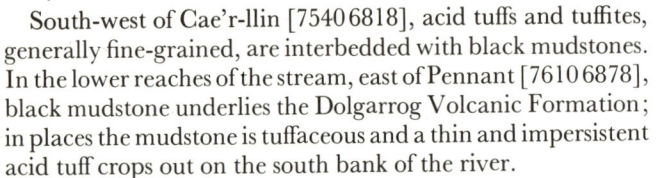

South-west of Cae'r-llin [7540 6818], acid tuffs and tuffites, generally fine-grained, are interbedded with black mudstones. In the lower reaches of the stream, east of Pennant [7610 6878], black mudstone underlies the Dolgarrog Volcanic Formation; in places the mudstone is tuffaceous and a thin and impersistent acid tuff crops out on the south bank of the river.

On Pen y gadair [739 693] the lowest rocks, dark grey mudstones with a fine-grained bedded basic tuff are poorly exposed. The overlying acid tuffs, up to 80 m thick, are rich in clasts, mainly of tubular pumice with some mudstone (E 39953, E 39961) and shards varying from large multicuspate to small fragmented spikes. The succeeding black and grey mudstones and siltstones on the western side of Pen y gadair include a thick acid tuff and tuffite; this dies out rapidly eastwards and the horizon is represented by tuffaceous siltstones with a thin tuffite.

Around Ochr Gaer [754 693], the sequence is highly faulted and intruded by both dolerites and rhyolites. The lithologies vary from clean vitric tuffs to grey and black mudstones. Admixtures of these two lithologies are common and vary from macro- to microscopic in scale. However, most exposures show evidence of disturbed or slumped beds with the pyroclastic fraction generally exceeding the sediment in amount.

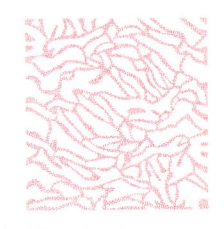

Farther south the formation occurs in the core of an anticline at Coed Dolgarrog [765 680] and Porth-llwyd. Here, evenly-

Cross-bedding

Tubular pumice

Vitroclastic texture in tuff

bedded acid tuffs and tuffites with black mudstones closely resemble the Middle Crafnant Volcanic Formation of the Sarnau area (Howells and others, 1978) and contrast with the slumped and highly mixed sequences in the vicinity. Admixtures of tuff and mudstone are common, with the mudstone in some instances (E41097) indicating possible incorporation in the tuff as unlithified clasts. Included in the sequence are isolated thin basic tuffs.

Interpretation The Middle Crafnant Volcanic Formation in the south of the district accumulated in a sedimentary environment of mudstone deposition. Incursions of thin ash-flow tuffs and banded fine-grained tuffites, probably air-fall in part, seem to have settled through the water column and caused no change in the sedimentation. Because the overlying Upper Crafnant Volcanic Formation is an ill-sorted mixture of pyroclastic material and mud comparable with the Upper Rhyolitic Tuff Formation of Dolwyddelan (Howells and others, 1973), establishment of a deep-water environment is assumed, though the deepening was probably earlier at Dolgarrog than at Dolwyddelan, judging from the intercalated and overlying black slates.

As in the ground north-east of Capel Curig, the undivided Middle and Upper Crafnant Volcanic formations, with disturbed and mixed lithologies, are interpreted as lying at the margin of an apron of slumped material derived from pyroclastics and mudstone emplaced and deposited to the north. Local eruptions of basaltic material occurred, as north of Llyn Cowlyd and in Afon Dulyn, though these also may have been disturbed and possibly comminuted.

In the northern part of the district, north of Afon Porth-llwyd, a marked thickening of the combined Middle and Upper Crafnant Volcanic formations to some 1000 m suggests a source nearby.

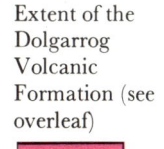

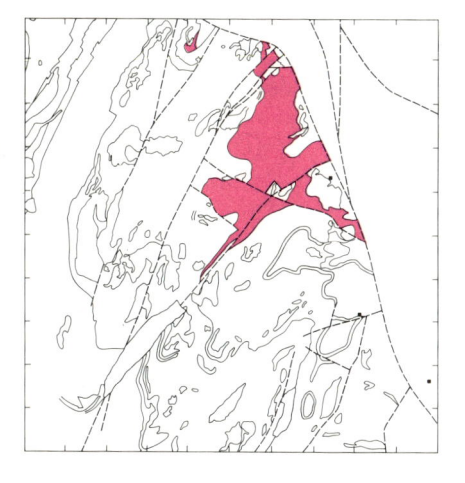

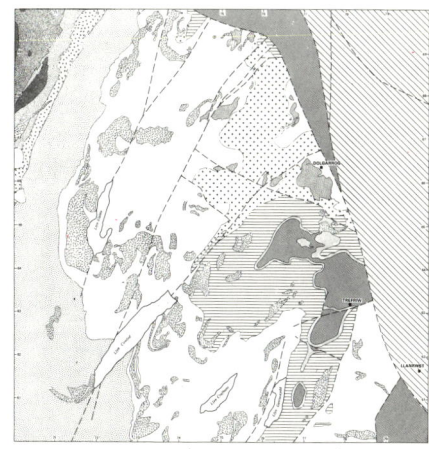

Extent of the Dolgarrog Volcanic Formation (see overleaf)

Dolgarrog Volcanic Formation

Around Dolgarrog and Penardda and in the ground to the north the Crafnant Volcanic Group is succeeded by a local pile of basaltic rocks, here named the Dolgarrog Volcanic Formation. Both Davies (1936) and Stevenson (1971) included these rocks in the Crafnant Volcanic Group.

The formation comprises some 300 m of basaltic rocks, classified mainly as hyaloclastites (see below), though there are also some crystalline basalts and basaltic tuffs, which are locally aggraded; no persistent mappable subdivisions have been found. Many of the rocks are massive, though there is a stratification in places. The most common lithology consists of feldspar crystals, fragments of vesiculated chloritised basaltic glass, and both whole and fragmented basalt pillows set in a comminuted and vesiculated chloritised glassy matrix. The pillows indicate subaqueous emplacement of basaltic magma, with glass fragments formed by the disintegration of the rapidly-chilled surface of the mobile lava at its contact with water.

The term hyaloclastite was proposed by Rittmann (1962) to describe the deposits formed by the comminution of glassy shells of growing pillows. However, Tazieff (1974) has pointed out that hyaloclastite deposits of this type are less abundant than those produced by the ejection of lava fragments into water. Silvestri (1963) suggested a genetic classification to extend from initial hyaloclastite breccias, formed locally immediately after the opening of the fissure source, to reworked hyaloclastites which could contain sedimentary material at a distance from the eruptive source. In the Dolgarrog Volcanic Formation the predominant rock type is hyaloclastite breccia on the Silvestri classification. Where the material is clearly reworked, comminuted and bedded by a sedimentary process, the term basaltic tuff is used.

Near Cae Coch Mine, basic tuffs and hyaloclastites of the formation are faulted against Llanrhychwyn Slates. The volcanic rocks vary from massive to well bedded and from coarse to fine-grained. The predominant lithology shows abundant idiomorphic feldspar crystals, up to 5 mm in length (E 40587) in a groundmass of chloritised vesiculated basaltic glass with much ilmenite. The feldspars are replaced by sericite, carbonate and quartz, and the vesicles filled with quartz and carbonate. Devitrification and recrystallisation have commonly obscured the fragmentary character of the glassy matrix, though elsewhere (E 40591) it accentuates the clastic character. Common rounded basaltic clasts, up to 8 mm in diameter (E 40593) bear feldspar microlites set in a glassy groundmass crowded with ilmenite grains. In places (E 40602) the rock is predominantly a glassy basalt with feldspar microlites, phenocrysts of sericitised albitised feldspar and some vesiculated glass fragments. Locally [7763 6556] the rock is thinly bedded in units up to 20 cm thick, the thicker beds being graded. Thin sections (E 40601) show feldspar crystals, glass with vesicles filled with spherulitic quartz, carbonate, micaceous material and prismatic crystals of epidote. Elsewhere finer bands of basaltic tuff with fewer feldspar crystals are relatively rare and indicate some reworking of the hyaloclastite pile.

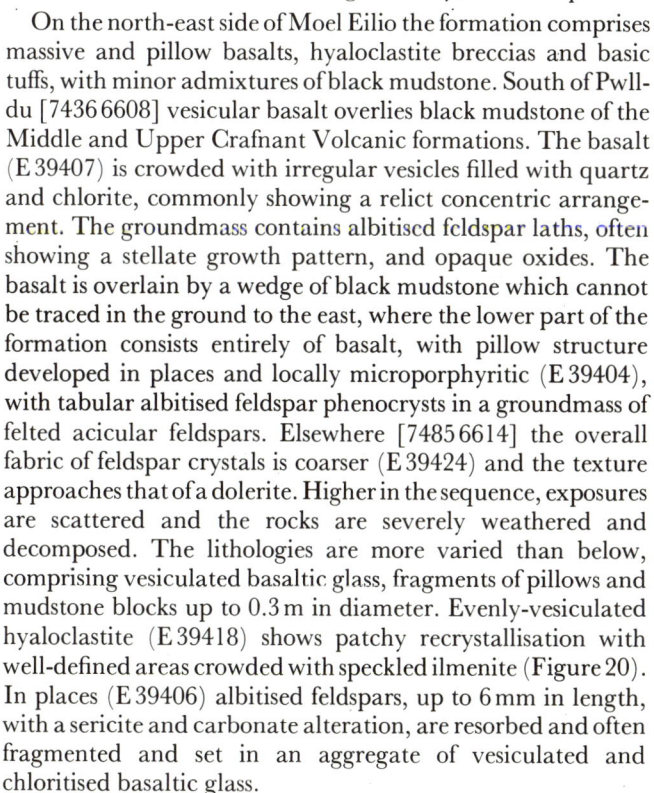

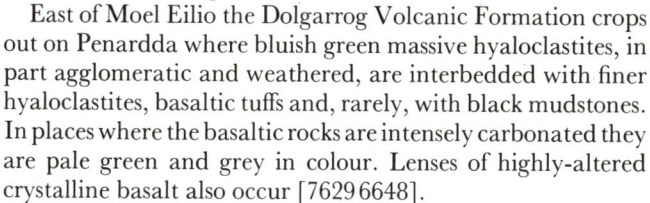

On the north-east side of Moel Eilio the formation comprises massive and pillow basalts, hyaloclastite breccias and basic tuffs, with minor admixtures of black mudstone. South of Pwll-du [7436 6608] vesicular basalt overlies black mudstone of the Middle and Upper Crafnant Volcanic formations. The basalt (E 39407) is crowded with irregular vesicles filled with quartz and chlorite, commonly showing a relict concentric arrangement. The groundmass contains albitised feldspar laths, often showing a stellate growth pattern, and opaque oxides. The basalt is overlain by a wedge of black mudstone which cannot be traced in the ground to the east, where the lower part of the formation consists entirely of basalt, with pillow structure developed in places and locally microporphyritic (E 39404), with tabular albitised feldspar phenocrysts in a groundmass of felted acicular feldspars. Elsewhere [7485 6614] the overall fabric of feldspar crystals is coarser (E 39424) and the texture approaches that of a dolerite. Higher in the sequence, exposures are scattered and the rocks are severely weathered and decomposed. The lithologies are more varied than below, comprising vesiculated basaltic glass, fragments of pillows and mudstone blocks up to 0.3 m in diameter. Evenly-vesiculated hyaloclastite (E 39418) shows patchy recrystallisation with well-defined areas crowded with speckled ilmenite (Figure 20). In places (E 39406) albitised feldspars, up to 6 mm in length, with a sericite and carbonate alteration, are resorbed and often fragmented and set in an aggregate of vesiculated and chloritised basaltic glass.

East of Moel Eilio the Dolgarrog Volcanic Formation crops out on Penardda where bluish green massive hyaloclastites, in part agglomeratic and weathered, are interbedded with finer hyaloclastites, basaltic tuffs and, rarely, with black mudstones. In places where the basaltic rocks are intensely carbonated they are pale green and grey in colour. Lenses of highly-altered crystalline basalt also occur [7629 6648].

In the bed of Afon Porth-llwyd [7590 6717] a pillow breccia contains complete and fragmented pillows up to 0.5 m in length in a matrix of chloritised and carbonated basaltic glass

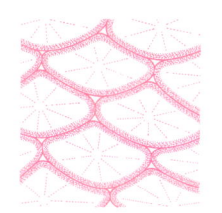

Pillow structures in basalt

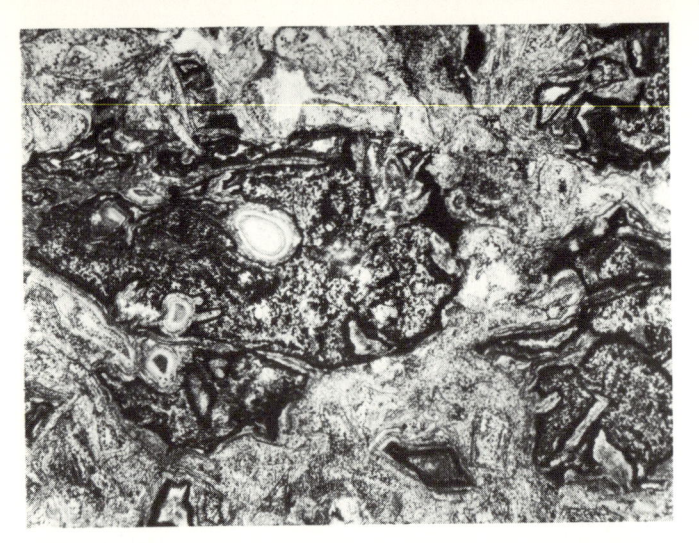

Figure 20 Hyaloclastite. Fragments of iron-rich devitrified basaltic glass in a carbonate matrix, E 39418. Plane polarised light, ×40

Figure 21 Pillow breccia, Dolgarrog Volcanic Formation. Fragmented basaltic pillows and lava toes in a hyaloclastite matrix. Pont Newydd, Afon Porth-llwyd

fragments (Figure 21). Albitised feldspar crystals are few, while leucoxenised ilmenite and sphene are common. Rounded basaltic lapilli up to 1 cm in diameter occur, with a crystalline fabric which is in sharp contrast to the vesiculated glassy host. The pillows are spilitic and consist of a felted mass of feldspar microlites showing a stellate arrangement with some interstitial chlorite, ilmenite and feldspar. Isolated feldspar phenocrysts, up to 2.1 mm long, are pseudomorphed by sericite and quartz and vesicles are filled with carbonate, quartz and some chlorite and surrounded by a concentration of opaque oxides.

Both the Dolgarrog and the underlying Middle and Upper Crafnant Volcanic formations are intruded by an irregular transgressive dolerite west of the Aluminium Works. Similar, although generally smaller, bodies have been recognised within the hyaloclastites in Afon Porth-llwyd and near Tan'r-allt, west of Tal-y-bont. It is possible that these represent feeders to the basic extrusives (Stevenson, 1971, p. 54).

North of Afon Dulyn, the formation is exposed in faulted blocks in the small scarp feature near Cae-Asaph where it is generally massive and comprises mainly crystal-rich hyaloclastites, although in places [7590 6915] lithic fragments and a mudstone matrix have been found and elsewhere [7583 6899] porphyritic basalt masses occur. In places individual, formerly glassy, fragments of the hyaloclastite matrix are indistinct. The matrix is chloritised, but in places (E41131) also shows replacement by a microcrystalline mosaic of feldspar, chlorite and opaque minerals. Some vesicles are deflected around resorbed and fragmented crystals of slightly altered plagioclase (E41136). Occasionally plagioclase crystals show a marginal replacement by potash-feldspar together with a little quartz and chlorite (E41132), the matrix of the rock being dominantly potash-feldspar. The basalts are porphyritic (E41137) with highly resorbed tabular albite-oligoclase phenocrysts, up to 2.4 mm in length, set in a felted mass of albitised plagioclase laths. Sphene occurs in isolated sub-idiomorphic crystals, up to 1.8 mm, often in association with chlorite, which occurs interstitially throughout the fabric.

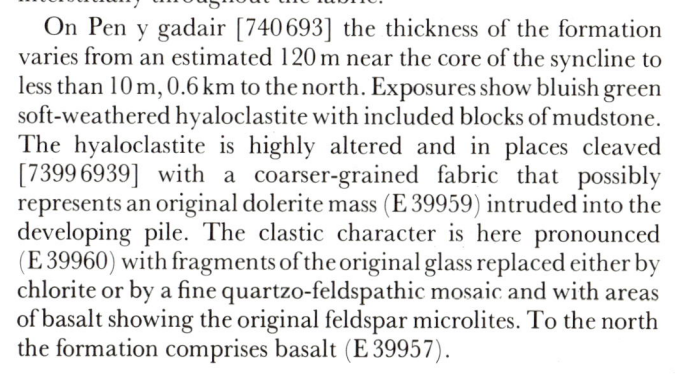

On Pen y gadair [740 693] the thickness of the formation varies from an estimated 120 m near the core of the syncline to less than 10 m, 0.6 km to the north. Exposures show bluish green soft-weathered hyaloclastite with included blocks of mudstone. The hyaloclastite is highly altered and in places cleaved [7399 6939] with a coarser-grained fabric that possibly represents an original dolerite mass (E 39959) intruded into the developing pile. The clastic character is here pronounced (E 39960) with fragments of the original glass replaced either by chlorite or by a fine quartzo-feldspathic mosaic and with areas of basalt showing the original feldspar microlites. To the north the formation comprises basalt (E 39957).

Interpretation The Dolgarrog Volcanic Formation accumulated as a result of submarine basaltic eruptions. The small transgressive bodies of dolerite at the base and within the formation probably represent some of the feeders to these eruptions. It is apparent from the included pillows that the eruptions were submarine (see p. 42), though the relatively small proportion of whole and fragmented pillows suggests that the major factor in the production of glass fragments was the expansion of entrapped steam within the erupted fragments. This process is repetitive as each fragmentation exposes fresh glowing surfaces and if the erupting source is far enough below sea level the original fragments will be completely pulverised. Continuing basaltic eruptions added to the relatively unstable pile and intrusions into the earlier accumulated material

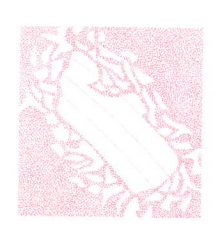

Resorbed sodic plagioclase

disturbed the equilibrium of the steep margins, resulting in their collapse and the development of a secondary brecciating process.

The limited lateral extent of the Dolgarrog Volcanic Formation indicates its initial accumulation in a depth of water sufficient to avoid emergence as there is little evidence of reworking of the volcanics in the overlying Llanrhychwyn Slates. This accords with the progressive deepening of the sedimentary environment throughout the period of the accumulation of the Snowdon/Crafnant Volcanic Group in the area of eastern Snowdonia, inferred on other grounds (p. 41). However, the depth of the water column was not enough to suppress the release of gases dissolved in the magma and inhibit explosive activity. Local development of bedded basaltic tuffs within the volcanics suggests current action around the volcanic pile. Local currents might be produced by heat convection above the volcanics or by the obstruction of deep-water currents by the pile. The formation of turbidity currents by collapse of parts of the rapidly forming hyaloclastite edifice might be expected and would explain the uniform graded bedding observed in places.

Graptolite *Orthograptus truncatus* Twice life size

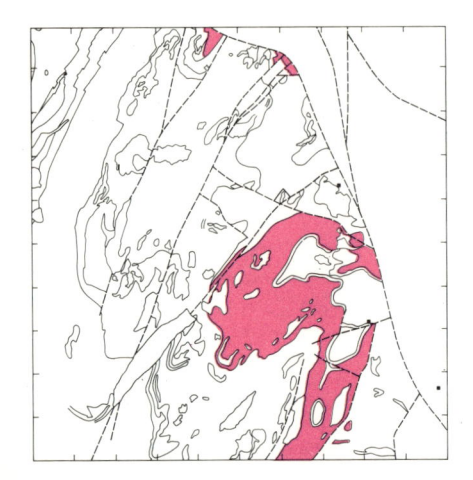

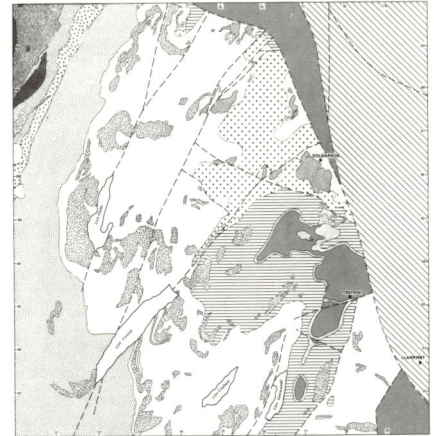

Llanrhychwyn Slates   6

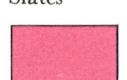

The Llanrhychwyn Slates, which are predominantly black, pyritic, well cleaved and up to 450 m thick, overlie the Crafnant Volcanic Group in the south of the district. To the north the Llanrhychwyn Slates overlie at least the peripheral parts of the volcanic pile of the Dolgarrog Volcanic Formation. They thin towards Dolgarrog but the nature of their relationship to the whole of the volcanics above Dolgarrog is unknown.

Stevenson (1971), following Wood and Harper (1962), placed the greater part of the Llanrhychwyn Slates in the *Dicranograptus clingani* Zone. In the Capel Curig district Howells and others (1978) recorded a *Diplograptus multidens* Zone fauna from the Llanrhychwyn Slates and the stratigraphically equivalent Black Slates of Dolwyddelan although recently Dr B. Rickards has re-examined the faunas collected by Williams and Bulman (1931) from the latter formation and has supported their determination of the *D. clingani* Zone. In the present district the Llanrhychwyn Slates have yielded a *D. multidens* Zone fauna at one locality, low in the sequence.

South of Pen-y-Drum the slates are exposed in a large number of trial excavations east of Tal-y-llyn. Patches of brecciated slate with pyrite mineralisation occur near the south end of Llyn Geirionydd [7635 6041].

South of Llanrhychwyn, the slates have been quarried below a dolerite sill at Pen-y-ffridd, where [7745 6092] (loc. 4, p. 85) they bear a graptolite fauna strongly suggestive of the upper part of the *D. multidens* Zone. Davies (1936) collected an exclusively biserial assemblage, which he ascribed to the zone of *Climacograptus wilsoni*, from this locality.

In a road-cutting [7806 6228] in Coed yr Allt, the slates include thin bands of soft ferruginous carbonate concretions. Higher in the sequence a tuffaceous mudstone composed of indistinct chloritic material with highly altered small plagioclase crystals can be traced for a short distance to the edge of the forest.

North of the Crafnant valley the slates crop out in numerous small exposures and stream sections between Bronfelen and Gymannog. To the south-west of Lledwigan they are tuffaceous at the base and a few small exposures show that isolated lenses of muddy tuff are also present locally.

On Cefn Cyfarwydd, exposures are few except in the vicinity of the dolerite intrusions near Brwynog-isaf, on the south side of

Graptolite *Diplograptus multidens* Twice life size

Extent of the Llanrhychwyn Slates

Afon Ddu. At Cae Coch Mine [7746 6544] the Llanrhychwyn Slates are faulted against basic tuff of the Dolgarrog Volcanic Formation. Black slates with pyritic spots crop out in Afon Ddu south-west of Tyddyn-Wilym and in trial levels farther north.

South of Llanbedr-y-cennin black pyritic slates with thin local interbedded basic tuffs are poorly exposed in Afon Dulyn south of Ffynnon-Bedr (loc. 5, p. 85), and in small faulted areas around Tyddyn-y-coed.

Interpretation Beds of similar lithology to the Llanrhychwyn Slates extend over a wide area of North Wales (Cave, 1965). Williams and Bulman (1931) assumed that the Black Slates of Dolwyddelan were deposited in shallow water "because they conformably overlie shallow-water admixed lithologies of the 'Upper Rhyolite Series'". Howells and others (1978) argued that there was no evidence to suggest that the Upper Rhyolitic Tuff/Upper Crafnant Volcanic formations were emplaced in shallow water and that the deep-water black slate environment, established within the Middle Crafnant Volcanic Formation, continued throughout the emplacement of the Upper formation. This conclusion is supported in the present district where there is little evidence in the Llanrhychwyn Slates of sediments resulting from the reworking of the underlying Upper Crafnant and Dolgarrog Volcanic formations.

Figure 22 Llanrhychwyn Church

7 Trefriw Mudstone and Trefriw Tuff

South of the Crafnant valley the Llanrhychwyn Slates are overlain by the Trefriw Mudstone which consists of up to 16 m of greyish black mudstones, with impersistent sandstone intercalations. Northwards there is a lateral passage into up to 72 m of pumiceous tuff and agglomerate (the Trefriw Tuff) which extends from Afon Crafnant, west of Trefriw, across Cefn Cyfarwydd to the vicinity of Cae Coch. Still farther north, near Llanbedr-y-cennin, the Trefriw Tuff/Mudstone horizon cannot be traced.

The Trefriw Mudstone crops out as outliers in the Pen y Drum Syncline where it includes some siltstone and fine sandstone bands. In the outlier east of Llanrhychwyn, parts of the sequence are sandy and contain pumiceous clasts which are prominent on weathered surfaces.

Near Trefriw the Trefriw Mudstone crops out in Coed y Gwmmanog as silty mudstone which is difficult in places to distinguish from the underlying Llanrhychwyn Slates. Farther east the sandstone element increases progressively on the eastern side of the syncline at Grinllwm, and in the crags of Coed y Wern the unit consists predominantly of up to 22 m of sandstone with a median band of cleaved silty mudstone up to 5 m thick. The sandstones are evenly bedded with pebbly layers, channel and scour features and bands showing slumped and contorted bedding. In the Afon Crafnant even bedding is seen in hard compact sandy tuff. On the south side of the valley [7793 6304] a highly-weathered basic tuff crops out. From Coed Creigiau [7740 6355] to approximately the highest part of Cefn Cyfarwydd basaltic tuffs of sandstone grade with agglomeratic bands are interbedded with siltstones. The basaltic tuffs show even flaggy bedding with cross lamination and consist (E 39741) predominantly of rounded clasts of vesicular basaltic glass, replaced by chlorite, with a few included albitised plagioclase phenocrysts. The coarse bands are generally more massive with rounded vesicular basaltic blocks, up to 0.3 m in diameter, crowded in a matrix (E 39742) of comminuted basaltic material.

The tuffs are exposed in small structural inliers on the steep scarp of the Conwy valley at Berth-lŵyd [7756 6480] and south of Ffrîdd-uchaf [7695 6495]. The general lithologies are here similar to those farther south with an admixture of flow-banded basalt and hyaloclastite within the tuffaceous matrix (E 39743).

The basaltic glass fragments are replaced by aggregates of green chlorite and the vesicles are filled with a feldspar mosaic with some included quartz, chlorite and carbonate.

In the vicinity of Rhibo [7735 6510] the Trefriw Tuff is intruded by dolerites and in a number of exposures it is difficult to distinguish the weathered dolerite from the basic tuffs. Davies (1936) considered that possibly 'an autobrecciated or pillowy flow' occurred near the top of the Trefriw Tuff at this locality.

Interpretation The lithologies and sedimentary structures of the Trefriw Tuff indicate reworking of basic volcanics in shallow water. The facies variation suggests that the source was in the Dolgarrog area. Probably the tuff results from a brief resurgence of activity from the volcanic centre which had given rise to the Dolgarrog Volcanic Formation. Alternatively it is possible that with relative uplift the upper parts of the volcanic pile were simply reworked in shallow marine conditions.

Extent of the
Trefriw Mudstone
and Trefriw Tuff

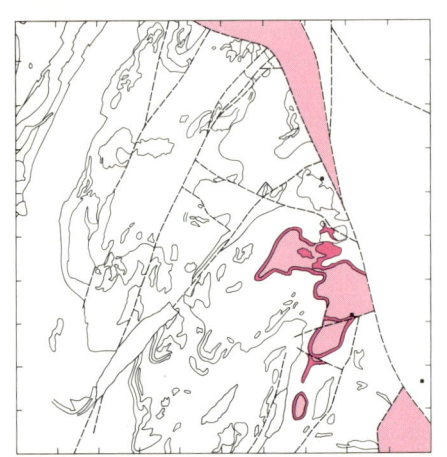

Extent of the
Grinllwm Slates

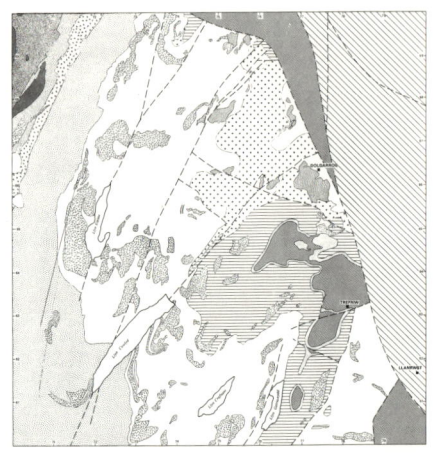

Grinllwm Slates

8

The Grinllwm Slates, up to 320 m thick, are grey mudstones, containing irregular thin bands and lenses of sandstone in the south though these are less persistent than in the Capel Curig district (Howells and others, 1978). In the Conwy area, to the north, the equivalent beds were divided into Bodeidda Mudstones and Deganwy Mudstones by Elles (1909) on faunal grounds. Lithologically, however, they are not separable. At Conwy the mudstones yield a shelly fauna allowing the lowest part to be tentatively placed in the Caradoc while the remainder is of Ashgill age.

South of Llanrwst, the escarpment above Gwydir uchaf is composed of grey slates and siltstones with flaggy bands, up to 8 cm thick, of cross-bedded sandstone showing ripple-marked bedding planes [7950 6083] and with lenses of highly-weathered carbonate and pyrite. In the outliers at Grinllwm (Figure 23) and Pen y Drum the formation consists of grey

Figure 23
Grinllwm, viewed from the south, showing eroded scarp features in the Grinllwm Slates

cleaved silty mudstones with little evidence of the sandstone ribs so characteristic of the sequence south of Llanwrst. Where the slates overlie the mudstone facies of the Trefriw Mudstone, the junction between the two is indistinct.

North-west of Trefriw the Grinllwm Slates crop out in Coed Creigiau, in numerous exposures on the slope between the village and Bryn-pyll, and to the south of Ffrîdd-uchaf. The lithology is predominantly a cleaved grey silty mudstone with infrequent bands of fine sandstone. Between Tal-y-bont and Llanbedr-y-cennin, numerous small exposures of grey silty mudstone show bioturbation structures. Stevenson (1971) records the presence of a poor shelly fauna at two localities [7530 6993 and 7531 6992] near Llanbedr-y-cennin.

Interpretation Shallow-water conditions, already established (see p. 50), continued during the deposition of the Grinllwm Slates. This is confirmed by sedimentary structures in the sandstones and siltstones and by the presence of the shelly fauna.

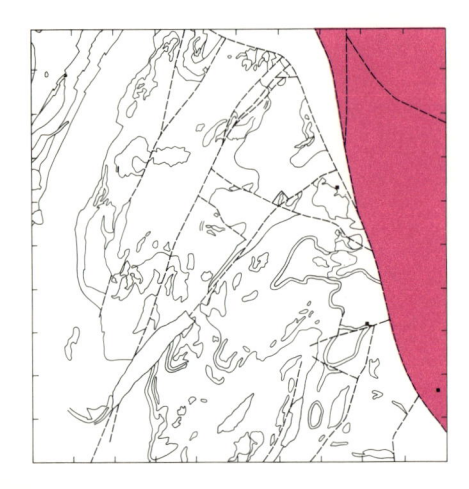

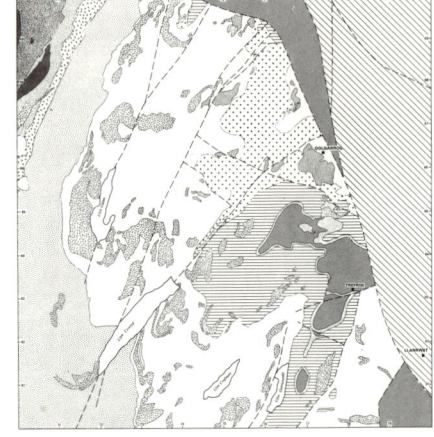

Silurian   9

Graptolite
Monoclimacis flumendosae   Twice life size

Extent of the Silurian rocks

The Silurian rocks of the ground to the east of the River Conwy form part of the large tract extending eastwards to the Vale of Clwyd. A full account of these rocks, including those of the present area, is given by Warren and others (*in press*).

The oldest Silurian rocks in the Dolgarrog area—the Upper Llandovery Pale Slates—occur beneath the alluvium of the Conwy valley. They are exposed [771 703] just to the north of the sheet boundary and consist mainly of pale greenish grey mudstones, locally mottled, which weather pale buff. Dark grey to grey mudstones are interbedded towards the base and top. With an increase in dark grey bands the beds pass down into undivided Llandovery mudstones (cut out by faulting in the present district) and up into the Denbigh Grits Group, of Wenlock age. The Pale Slates range in age from the *Monograptus turriculatus* Zone to the top of the *Monoclimacis crenulata* Zone.

The Denbigh Grits Group forms the rising ground to the east of the Conwy valley. The total thickness is about 1100 m though only 750 m crop out within the present area, the lowest beds being probably faulted out throughout much of the area while the highest lie beyond the eastern margin. The group consists of sandstones, striped silty mudstones and disturbed beds, the proportions of which vary both vertically and laterally.

The sandstones, up to 75 m thick, vary from fine to coarse-grained, are locally pebbly and are of greywacke type (Pettijohn, 1957). Sole structures occur at a number of localities and indicate a general westerly source.

The striped silty mudstones, forming the bulk of the Denbigh Grits Group, comprise mudstones with bands of siltstone or fine sandstone. Thin bands of laminated muddy siltstone, calcareous siltstone and shelly bands with an allochthonous fauna also occur, the last two both weathering to a brown 'gingerbread' rock.

Disturbed beds are strata in which the bedding is highly contorted and/or fragmented and is in places completely obliterated. They range from less than 1 m to more than 80 m in thickness and many of the types recognised by Warren and others (*in press*) occur within the present district: (i) bedded, (ii) unbedded, (iii) pencil slate, (iv) pebbly/sandy mudstone and (v) sandstone disturbed beds. All the disturbed beds have basal contacts which are sharp or involve rapid transitions to undisturbed rocks. Upper contacts show sandstone or striped silty mudstone resting with marked 'unconformity' on the disturbed bed though the contact surface is often near-planar. The genesis of these beds was for long in dispute (see for example Boswell, 1953; Jones, 1954), but they are now generally considered to be the products of penecontemporaneous slumping or sliding on the sea floor, as maintained by Jones.

Graptolites, sparse elsewhere and rare in the present district, are the principal faunal element in the Denbigh Grits Group. The shelly fauna is largely restricted to broken fragments of shells in sandstone and to 'gingerbread' horizons in striped silty mudstone. The faunas obtained from the present district are listed on pp. 85–87 (locs. 6–8). In terms of graptolite zones the beds exposed appear to lie mainly within the zone of *Cyrtograptus rigidus* but probably extend down into the *Monograptus riccartonensis* Zone and up into the *Cyrtograptus perneri* Zone.

The succession includes four units of sandstones and disturbed beds (Figure 24). The lowest unit crops out near the River Conwy and is seen in the road-cutting [7813 6963] north of Dyto where the soles of some of the sandstones show linear load-casts. In a quarry [7821 7001] just north of the sheet, a disturbed bed cuts down into the underlying sandstone, one bed of which shows longitudinal ridge casts.

The second unit forms the conspicuous sandstone scarps north of Dyto, where bases of individual beds locally [7824 6932] show linear load-casts. Though correlation to the south is not certain the unit is probably represented solely by sandstone between the Dyto Fault and the wood [785 676] east of Bodhyfryd. Southwards to Plas Maenan [7870 6675] disturbed beds predominate.

The third unit consists of varying proportions of sandstone and disturbed beds. It is well seen in Penllyn Wood [785 696]. To the south of the Dyto Fault the probable correlative forms the prominent scarps [789 668] east of Plas Maenan, where there are sections in thick sandy disturbed beds and sandstones. Farther south disturbed beds, forming the upper part of the same unit, are seen in the road- and railway-cuttings at Plas Madoc Lodge [7932 6292], where they consist of sandy siltstone enclosing isolated sandstone and siltstone blocks (type iv). These beds die out [793 639] north of Tan-y-fron. At Tan-lan, a quarry [7915 6352] shows 10 m of an underlying massive sandstone with thin siltstone bands and lenticular conglomerates up to 30 cm thick overlain by 1.5 m of flaggy siltstone; casts of current lineations occur on the soles of the sandstone beds.

A typical section of the beds between the third and fourth units is exposed in the stream [7872 6916] west of Hafotty. It comprises 9 m of striped silty mudstones, with sporadic fine-grained laminated sandstone bands up to 12 cm thick and with groove and flute-casts together with interference ripple casts.

Graptolite
Cyrtograptus rigidus
Life size

Brachiopod
Rhynchotreta cuneata
$1\frac{1}{2}$ times life size

The fourth unit is thickest between Hafotty and Ty'n-y-bryn [7957 6852] where it consists predominantly of highly cleaved pencil slate disturbed beds (type iii), which locally contain pebbles up to 2.5 cm in diameter of vein quartz, silty mudstone and black mudstone. Associated fine to coarse-grained sandstones are interbedded with the disturbed strata. To the south of the Dyto Fault the unit is probably represented mainly by sandstone with disturbed beds developed around Maes-y-groes [7968 6719] and Tan-y-graig [7987 6631]. Farther south the sandstones tend to die out and the unit is represented by disturbed beds, with subsidiary sandstones [7948 6413] near Garthmyn. South [7966 6316] of Plas Madoc disturbed sandstones, siltstones and mudstones pass laterally into disturbed sandy siltstones containing rafts of sandstone and slump conglomerates (type v). At about the same horizon a small exposure [7959 6209] in Station Road, Llanrwst, shows crudely cleaved silty mudstone with a few small slump-folded sandstone lenses and fragments (type iii).

In places on the eastern margin of the area the lowest part of the highest disturbed bed in the Denbigh Grits Group crops out. This is composed almost entirely of pencil-slate disturbed beds (type iii) though it contains in places beds of coarse sandstone and conglomerate in the lower part.

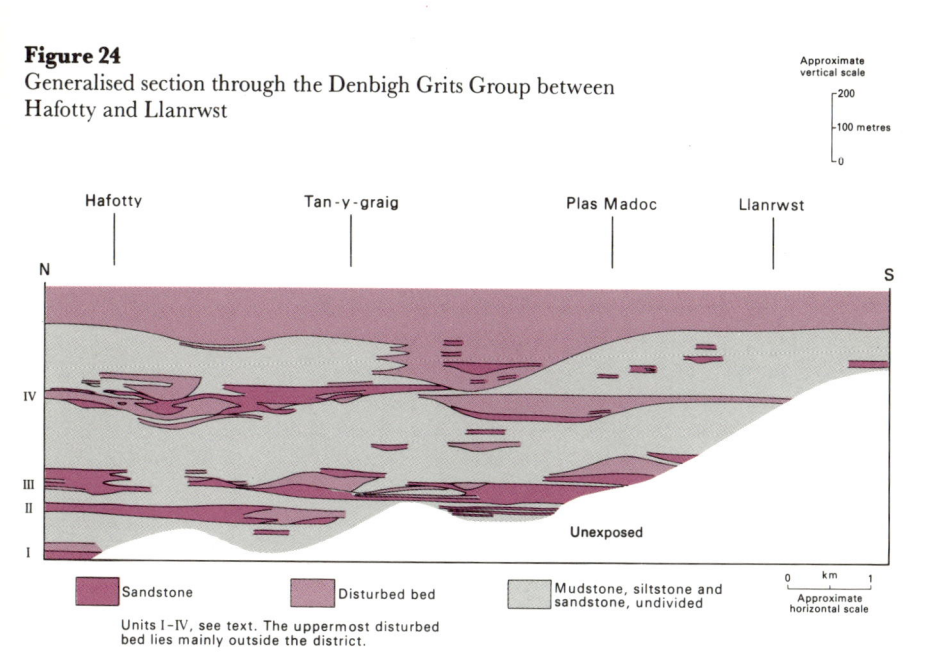

Figure 24
Generalised section through the Denbigh Grits Group between Hafotty and Llanrwst

Units I–IV, see text. The uppermost disturbed bed lies mainly outside the district.

Interpretation The areal distribution and lithology of the Silurian rocks in North Wales indicate that the present area lay within a basin of deposition which was elongated approximately east–west. It is believed that the basin was bounded to the west, along the line of the Conwy Valley Fault, by a land mass yielding recognisable Ordovician and Pre-Cambrian detritus. Sandstones in the Denbigh Grits Group appear to have been derived from this land mass and deposited as the products of near-shore slides and turbidity currents. Disturbed beds are interpreted as resulting from down-slope slumping, probably induced by movement and/or earthquake shocks along the line of the Conwy Valley Fault. Quieter conditions resulted in the deposition of the striped silty mudstones, probably on a gentle slope in the basin of deposition.

Figure 25
Pont Fawr, Llanrwst

Structure

10

The Lower Palaeozoic rocks of Wales were subjected to intense earth movements in the Caledonian Orogeny, which culminated in the late Silurian and early Devonian period. However, many of the structures within these rocks developed at an earlier stage and their evolution in the North Wales region has been discussed by Shackleton (1954) and Helm and others (1963). The structural history as described for the Capel Curig district (Howells and others, 1978) is equally applicable to the Dolgarrog district.

The district is divided structurally by the Conwy Valley Fault. To the east the Silurian shows an evenly-bedded sequence, dipping gently, for the most part eastwards, with an intricate pattern of faults, generally of small displacement and some minor folding. To the west of the Conwy valley, the Ordovician is affected by a complex pattern of interacting folds, whose styles tend to vary with the lithology. Faulting is less reticulate in type than that in the Silurian.

Folds

The district lies within the Snowdon Synclinorium which has a general slight north-easterly plunge. Trends of the major folds (Figure 26) vary from north-north-easterly in the south part of the district to east-north-easterly and easterly in the area adjacent to the Conwy valley. The folds range from gentle to isoclinal, with axial planes generally showing a steep inclination to the north-west. There is little evidence of hinge zone thickening in mesoscopic folds.

In the folds at the southern margin of the district there is a wide range in the degree of closure. The Pen y Drum Syncline, in rocks high in the sequence, is an open structure. Farther west, in the Crafnant valley, folds in rocks lower in the sequence are tighter with steeply inclined and overturned east-facing limbs and common gently inclined west-facing limbs. All these folds show a marked tightening around their northerly closures where their axial-plane traces swing to a more easterly trend.

A dominant easterly trend of axial-plane traces is apparent in the ground near the western side of the Conwy valley, particularly in rocks above the Crafnant Volcanic Group. A complex fold pattern occurs between Dolgarrog and Trefriw. The folds tend to be tight and have steeply inclined axial planes while in some instances the southern limbs are overturned.

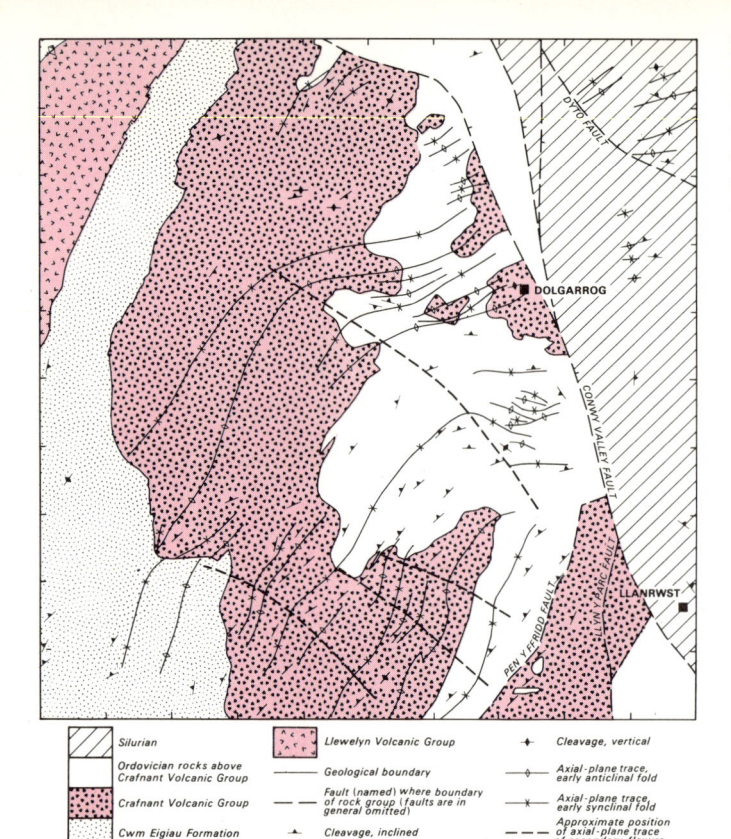

Figure 26
Sketch map showing the distribution and orientation of structures in the district

Variations in plunge and the deep dissection of the Conwy valley cause the Crafnant Volcanic Group to crop out at the bottom of the steep scarp in the Dolgarrog area.

The presence of later folds, trending approximately north-west, can be inferred from flexures in the early major fold axes and associated cleavage. These later flexures are similar to those described in the Capel Curig district (Howells and others, 1978). Although early folds were probably tightened during this later deformation phase, their distinctive plunge culminations and depressions are probably the result of initial inhomogeneous strain.

A number of NE to E-trending folds are to be found in the Silurian rocks, particularly to the north of the Dyto Fault. These folds are in general symmetrical, with vertical or steeply north-dipping axial planes, and all plunge gently eastwards. By Cefnycoedisaf [7871 6835], an anticline and complementary syncline are present, both trending and plunging north. Elsewhere it can be shown (SH 77, also Warren and others, *in press*) that N–S folds precede easterly-trending folds.

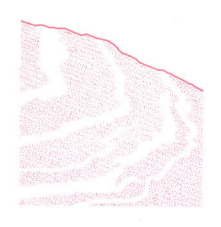

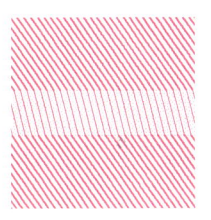

Cleavage

The cleavage is axial planar to the fold structures and its intensity is closely related to lithology. In the mudstones it is slaty, with platy minerals showing a high degree of preferred orientation. In the coarser greywacke-type sandstones recrystallisation of quartz and micaceous minerals in the matrix has produced a strong preferred orientation with parallel intergranular fracturing. In the finer sandstones the matrix generally forms a smaller proportion and as a result preferred orientation fabrics are less apparent. In such sandstones the fractures are commonly concentrated in widely spaced zones.

The volcanic rocks are generally poorly cleaved, except where there is a fine matrix or, as in the basal zones of the thick ash flows of the Lower Crafnant Volcanic Formation, a high micaceous content. In the highly siliceous tuffs there is little sign of a tectonic fabric.

In the slates associated with the tight folds of the Crafnant valley, for example in the slates underlying the Upper Crafnant Volcanic Formation on Mynydd Deulyn and in the Llanrhychwyn Slates, mineral growth along cleavage is indicated by elongated pyrite spotting and strain shadows on small scattered tuffaceous inclusions. Generally such growths are ill defined although all clasts and nodules in tuffs and fossils in sediments show varying degrees of deformation. The long axes of mineral growth, where discernible, plunge at a high angle on the cleavage planes and are comparable with those described from the Dolwyddelan Syncline (Howells and others, 1978).

Apart from bedding, fracture cleavage is the most important planar structure in the Silurian rocks. Its development is sporadic whilst its intensity varies with lithology in any one area. In general the cleavage is vertical or dips steeply northwards and has an E–W orientation. Other trends also exist and include a N–S direction which outside the district has been shown to precede the main E–W cleavage (Warren and others, *in press*). 'Cleavages', often folded, in disturbed beds are excluded here as these fractures probably resulted from intense penecontemporaneous deformation.

Faults

Faults in the Ordovician are generally steeply inclined normal faults showing no signs of lateral movement. The fault pattern shows no fixed relationship to the main Caledonian folds. Most of the faults fall into two groups, NNE–NE and NW–WNW. The Conwy Valley Fault also lies between these two main trends for the main part of its course within the district and there is a tendency for other faults of E–W trend to be developed in the vicinity.

The prominent alignments of the Crafnant and Cowlyd valleys and Cwm Eigiau appear to be in part fault-controlled, although little displacement occurs along the faults. Similarly, in the Llanrhychwyn area the north-north-easterly Pen y ffridd Fault and the west-north-westerly fault on the south side of Grinllwm show good surface features although in both instances displacement is relatively small. By contrast the inclined fault extending from Llyn Eigiau to Llyn Cowlyd is little reflected by surface features for most of its course, although its displacement reaches about 360 m in places.

The Conwy Valley Fault is the most important fracture within the district. It separates not only Silurian rocks to the east from Ordovician rocks to the west, but also areas of different tectonic styles. Trending north-north-west, the fault lies entirely beneath superficial deposits within the Dolgarrog district and its throw must be of the order of 1800 m in the vicinity of Dolgarrog.

Although it is represented on the map as a single fault over much of its length, the Conwy Valley Fault is probably complex. North of Dolgarrog it divides into three branches. One branch follows the valley terminating the WNW–NNW fractures in the Silurian rocks to the east. A second branch leaves the first and separates Silurian from Ordovician rocks along the lower reaches of the Afon Roe. The third branch, with perhaps the greatest movement and representing the main fault, extends WNW towards Llanbedr-y-cennin. South from Llanrwst and outside the area of the map the fault again splits into a number of branches and there is evidence for sinistral wrench movement (Warren and others, *in press*).

The Conwy Valley Fault is a major, deep-seated fracture which is believed to have influenced sedimentation in Silurian (p. 56) and possibly in Ordovician times. It was active over a long period, the last movement post-dating the lead-zinc mineralisation of the Llanrwst mining district, which is believed to have continued into the Permian period after an initial emplacement in Viséan-Namurian times (Ineson and Mitchell, 1975, p. 89).

The Silurian rocks are affected by three distinct groups of faults of NNW, WNW and NE trends. NNW faults downthrow to either east or west, but outside the district they can be shown to possess lateral movement also (Warren and others, *in press*). WNW faults generally downthrow to the north-east and a major fault of this kind is the Dyto Fault. NE faults are less common but occur sporadically throughout the district. On the basis of their traces these faults, like the other two fault groups, are interpreted as being high-angle fractures.

Mineralisation   11

The area includes the northern part of the Llanrwst mining district which was once one of the most important sources of lead and zinc in the Lower Palaeozoic rocks of North Wales. The mining district is bounded by the Conwy valley to the east, and the Geirionydd valley to the west and it extends north to include a few small mines in the Llanrhychwyn Slates on the north side of the Crafnant valley. In addition, the Cae Coch Mine north of Trefriw, slightly outside the district defined above, is the only mine in North Wales to have been worked solely for pyrite.

The mineral potential of the district was recognised as early as 1625 (Dewey and Smith, 1922, p. 59), though early attempts to exploit the mineral resources were hindered by difficulties of extraction and transport. By the early 19th century access to railways was possible and provided a stimulus to mining activity, most intensive during the period from about 1850 to the outbreak of war in 1914. Mining declined as cheaper sources were discovered elsewhere and output dropped sharply. The Parc Mine veins, mainly outside the present district, were worked intermittently until the late 1950's (Dennison and Varvill, 1952; Archer, 1959).

The properties, mines and disposition of the main veins are shown in Figure 27. The Rhuad y rhwst Mine and part of the Parc and Hafna mines are situated in the Nant Gwydir valley. The Tan yr Eglwys and Pandora mines lie on the eastern limb of the Pen y Drum Syncline. The Tal y Llyn, Llyn Geirionydd, Glan Geirionydd and Klondyke mines are situated on the western flank of the anticline along the line of Llyn Geirionydd. Isolated properties of Ty'n twll and Coed Gwydir mines occur on the western scarp of the Conwy valley and Gellinewydd and Gellilydan mines in the Llanrhychwyn Slates on the north side of the Crafnant valley.

The following summary is based mainly on publications by Dewey and Smith (1922) and Archer (1959), supplemented by an unpublished report by Robertson made in 1940 and research by Marengwa (1973).

The mineralisation is of vein type, following steeply dipping normal faults which generally show little or no displacement. Dewey and Smith (1922) described three main trends, the earliest ENE set being displaced by N–S veins and both displaced by ESE veins. The N–S veins, as in the district to the south, tend to form broad mineralised belts generally

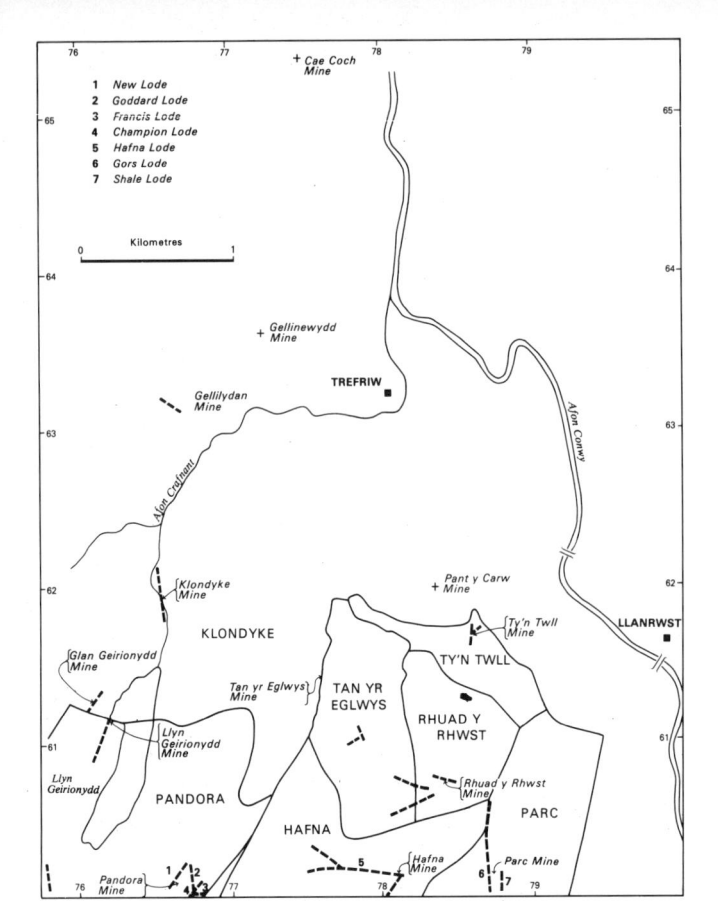

Figure 27
Sketch map showing lodes and properties of the northern part of the Llanrwst mining field

contrasting with the other narrow veins with well-defined walls. The nature and disposition of the veins is dependent on the lithology of the country rock. In many of the most important mines dolerite occurs in the vicinity of the lodes, sometimes forming the walls. Most of the intrusions are thus clearly older than the lodes but it is uncertain whether they exerted any control over the mineralisation.

The faults associated with the mineralisation post-date the cleavage and are probably of late Caledonian age. Ineson and Mitchell (1975) carried out age determinations on potassium-rich clay mineral assemblages in the veins and obtained evidence of an initial phase of mineralisation of Viséan–Namurian age and a succeeding phase of Permian age. The latter may correspond to brecciation of mineral deposits and subsequent mineralisation in some of the veins.

The predominant ore minerals are galena and sphalerite, although pyrite and marcasite are locally common. Chalcopyrite and magnetite have been identified in the Parc Mine (Archer, 1959) although they are rare. The main gangue minerals are quartz and calcite and their relative abundance is related both to their mineral association and to the lithology of the country rock in which the veins occur. Thus as a general rule breccias of slate are cemented by quartz whereas breccias of tuff are cemented by calcite.

Over the mining field as a whole it has been estimated (Dewey and Smith, 1922) that the proportion of ore to gangue averages from 8 to 15 per cent of galena and 8 to 10 per cent of sphalerite. Clearly the proportion of galena to sphalerite varies in different veins although details of the proportions from the old records are rare; at the Parc Mine the ratio was given as 2 to 1.

The main mines of the district are those situated in the Nant Gwydir valley. In the Dolgarrog district the main lodes of the Parc Mine are the Shale and the Gors, which run N–S, closely associated with the Llyn y Parc Fault. The records from the Parc Mine show that over the period from 1860 to 1913, 1140 tons of lead ore and 984 of zinc ore were extracted and from 1913 to 1930 approximately 1000 tons of galena concentrates were sold; these figures relate to the mine as a whole. From 1952 to 1954 Parc Mine produced 5176 tons of lead concentrates and 2146 tons of zinc concentrates (Thomas, 1961, p. 201).

On the western flank of the Nant Gwydir the veins worked in the Hafna and Rhuad y Rhwst mines are of predominant trend near E–W, with some NW and SW-trending cross-courses. Above the scarp of Clogwyn Gwlyb the lodes have been stoped to the surface, though the workings are now largely overgrown. The E–W vein at Hafna Mine, closely associated with a dolerite intrusion, is still well exposed in places. Between 1860 and 1913, 802 tons of lead ore and 2555 tons of zinc ore were extracted from the Hafna Mine although a proportion of these figures would have been obtained from the N–S vein extending beyond the southern limit of the area.

Of the remaining mines, the New Pandora and Klondyke are the most important. At the New Pandora, four principal veins, the New (NE–SW), the Goddard (N–S), the Champion (NE–SW) and the Francis (E–W), have been worked in the Dolgarrog district. Intersections of the veins have shown associated enrichment (Dewey and Smith, 1922). In the years from 1877 to 1911, 1377 tons of lead ore and 1939 tons of zinc were extracted.

The Klondyke Mine, in the valley of Afon Geirionydd, consists of levels and open workings in a broad nearly N–S belt of mineralisation in tuffs of the Upper Crafnant Volcanic Formation. Records of the mine are extremely poor.

Records of the remaining mines are sparse although in most cases development was small. Ty'n twll Mine, situated in the vicinity of a fault line, is one of the larger and it is recorded that between the years 1857 and 1875 the output of lead ore amounted to 372 tons which yielded 282 tons of lead, a part of which contained 5 oz of silver per ton (Dewey and Smith, 1922).

At Gellinewydd, west-north-west of Trefriw, a NW–SE vein, carrying pyrite associated with quartz and calcite, has been worked in three levels. A similar occurrence of pyrite at the same horizon has been worked at Cae Coch. Here the main mineralisation has been controlled by the fault which extends south-eastwards from east of Blaen-y-wern and throws down Llanrhychwyn Slates against the basaltic tuffs of the Dolgarrog Volcanic Formation. The fault zone and the adjacent Llanrhychwyn Slates are intruded by thin transgressive bodies of dolerite and larger dolerite intrusions occur in the vicinity on either side of the fault. Sherlock (1919) considered the orebody to be of sedimentary origin although subsequent unpublished reports by Robertson and Schnellmann indicate that it follows the contact of the Llanrhychwyn Slates with the dolerite and is also controlled by the structure. The epigenetic nature of the deposit is further confirmed by the occurrence of pyrite mineralisation in the rocks adjacent to the faults between Llyn Cowlyd and the Conwy valley.

The orebody, up to 2 m thick, consists of an admixture of pyrite and quartz with some calcite, and the adjacent dolerite and slates are pyritised. Archer (1959) records that approximately 90 000 tons of ore were extracted between 1860 and 1875 and 1600 tons were produced when the mine was reopened and worked by the government during the First World War.

Figure 28
Tailings from Parc Mine, recently levelled and cultivated

Intrusive igneous rocks

12

Intrusions of Lower Palaeozoic age occur throughout the Dolgarrog district and form an important component of the geology which influenced the structures produced by the late-stage Caledonian movements. They can be broadly divided into acid, intermediate and basic. The acid intrusions are rhyolites restricted to the north-eastern part of the district. Intermediate intrusions occur within the Foel Frâs Volcanic Complex in the north-west of the district, and are so closely related to the extrusives of the complex that they have been described together (p.10). The basic intrusions, dolerites with spilitic affinities in places, are the most extensive and occur mainly as sills throughout the district.

Acid intrusive rocks

Intrusive rhyolites occur around Dolgarrog and Llanbedr-y-cennin in the Middle and Upper Crafnant Volcanic formations and the Dolgarrog Volcanic Formation. The largest intrusion, near Dolgarrog, is over a kilometre in length, while smaller bodies occur to the south and west of Penardda, west of Porth-llŵyd and in faulted outcrops south of Llanbedr-y-cennin. The intrusions are usually of small lateral extent, an indication of the high viscosity of the magma, and tend to deform the adjacent strata. The intrusive rhyolites are folded and faulted in the same manner as the country rocks, which they barely post-date, and some degree of cleavage is commonly present.

The intrusions show varying degrees of concordance. For example, that west of Porth-llŵyd lies concordantly within a tightly-folded and evenly-bedded acid tuff and mudstone sequence, while the intrusion west of Dolgarrog transgresses the Middle and Upper Crafnant Volcanic formations to the base of the Dolgarrog Volcanic Formation.

In the larger bodies steeply dipping to vertical flow banding is common. Flow-brecciation is also a common feature, for example near the margin of the intrusion at Coed Dolgarrog [7695 6644] and similar brecciation occurs in restricted areas within the central part of the body. Xenoliths of country rock are present in places while siliceous nodules are more common in the smaller bodies, as on the south side of Penardda [7605 6615].

In hand specimen, the rhyolites are grey or greyish green, fine-grained and microporphyritic rocks, generally massive and well jointed. A perlitic structure occurs in places and is most easily recognised on weathered surfaces.

Perlitic texture in rhyolite

In thin section, textures of the rhyolites are usually obscured by crystallisation. In rare cases the original fabric can be observed, as at Clogwyn Mawr [7764 6564] (E 39966) where ghost-like remnants of feldspar microlites occur in recrystallised quartzose aggregate. Elsewhere, as at Porth-llŵyd [7672 6753] (E 41600), the microlites are better defined and in both instances rounded and resorbed tabular albite and potash feldspar phenocrysts, up to 2.2 mm, occur. In places (E 41597) remnant feldspar phenocrysts are all that remain of the original fabric. Spherulitic growths have been observed (E 40592) although only in isolated fragments within a cleaved recrystallised fabric.

Perlitic structure, an indication of the glassy character of the original material, is the most persistent feature to survive recrystallisation (E 40588) (Figure 29); the fractures are accentuated by segregations of chlorite and quartz and are clear in the hand specimen. In the intrusion to the south-west of Rowlyn-isaf [7520 6770] the perlitic rhyolite is brecciated, with an iron-rich silty mudstone matrix (E 41098) (Figure 30). The mudstone includes small slivers and tabular fragments aggraded from the perlitic rhyolite. The bedding within the mudstone and the form of the interfaces with the rhyolite indicates that it was caught up in the intrusion in an unlithified state. It is suggested that the intrusion extended to a high level, disturbing unlithified muds which became entrapped within the rapidly chilling margins of the rhyolite.

Basic intrusive rocks

The general characters of the intrusive dolerites of the Ordovician of North Wales were described by Harker (1889). As a result of their concentration within the Ordovician it was considered for many years that the intrusions were part of the

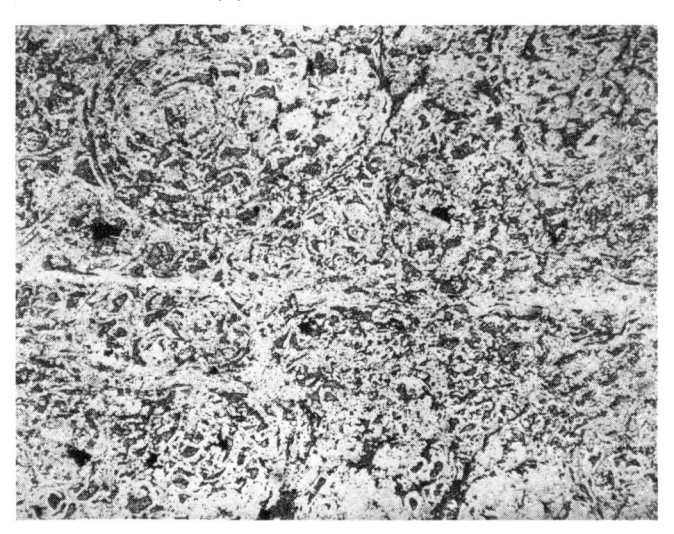

Figure 29 Perlitic-fractured rhyolite, E 40588. Plane polarised light, × 40

Figure 30
Brecciated perlitic-fractured rhyolite with iron-rich silty mudstone matrix, E 41098. Plane polarised light, × 40

Columnar jointing

Ordovician volcanism. Shackleton (1954) suggested that, although in certain instances this connection is valid, many were intruded at a late stage in the Caledonian movements.

In the Dolgarrog district the intrusions are mainly concentrated within the outcrop of the Crafnant Volcanic Group (**Figure 31**). At Bwlch Cowlyd sills intrude sandstones and mudstones below the Soudleyan–Longvillian boundary and at Llanrhychwyn, the Crafnant valley and Cae Coch they intrude the Llanrhychwyn Slates. The main intrusions form distinctive features on the ridges between Llyn Crafnant and Llyn Eigiau and the largest intrusion crops out in the steep crags of Craig Eigiau [7150 6530] and Cerrig Cochion [7230 6700].

There is evidence of folding later than the intrusions, as in the concordant sill-like body between the No. 1 and No. 2 tuffs of the Lower Crafnant Volcanic Formation. The thick sill on Pen y gadair is folded in a similar manner, following the synclinal structure of the surrounding tuffs. All dolerites occurring in the area have been affected to some extent by cleavage, especially at the margins. There is no evidence of any of the intrusions being controlled by faulting.

In all the well-exposed intrusions columnar jointing is common, for example at Marian Mawr. Where the contacts are exposed chilled marginal zones, of noticeably finer grain than the central parts, are rarely greater than 2 m thick. Similarly the zone of induration in the adjacent sediments is generally narrow with little indication of the proximity of the intrusion beyond 3 m.

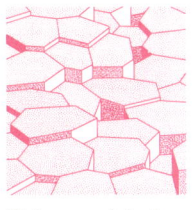

There is little petrographic variation in the composition of the dolerites, all having been albitised. The intrusion near the Hafna Mine in the Nant Gwydir valley shows plagioclase laths around An_{35} to An_{30} (E 40682, E 40685) in composition. In the Pen-y-ffrîdd dolerite, near Llanrhychwyn, the feldspar composition tends to be more sodic, in the range of An_{35} to An_{20}. The feldspars are superficially homogeneous but ghost-like remnants of a formerly more calcic plagioclase can be discerned. Secondary flakes of chlorite are developed along cleavage traces.

The pyroxene is ophitic with an overall prismatic habit. It is a slightly pleochroic, pinkish (titaniferous?) augite with common peripheral growths of tremolite-actinolite and some replacement by chlorite. Commonly the alteration is so intense that the pyroxenes are completely obliterated (E 40686).

Granular chlorite pseudomorphs after olivine, generally about 0.2 mm in diameter, are quite abundant. These pseudomorphs, unlike the pyroxenes, are determinable even in the most highly altered rocks. In places (E 40687) the olivine pseudomorphs tend to be grouped together in clusters and crystals range up to 1.5 mm in diameter. The peripheries and arcuate features of the crystals are accentuated by magnetite with the cores replaced by a pale green to colourless chlorite with anomalous brownish green birefringence.

Ophitic texture in dolerite

Figure 31 Sketch map showing distribution of dolerite intrusions at outcrop

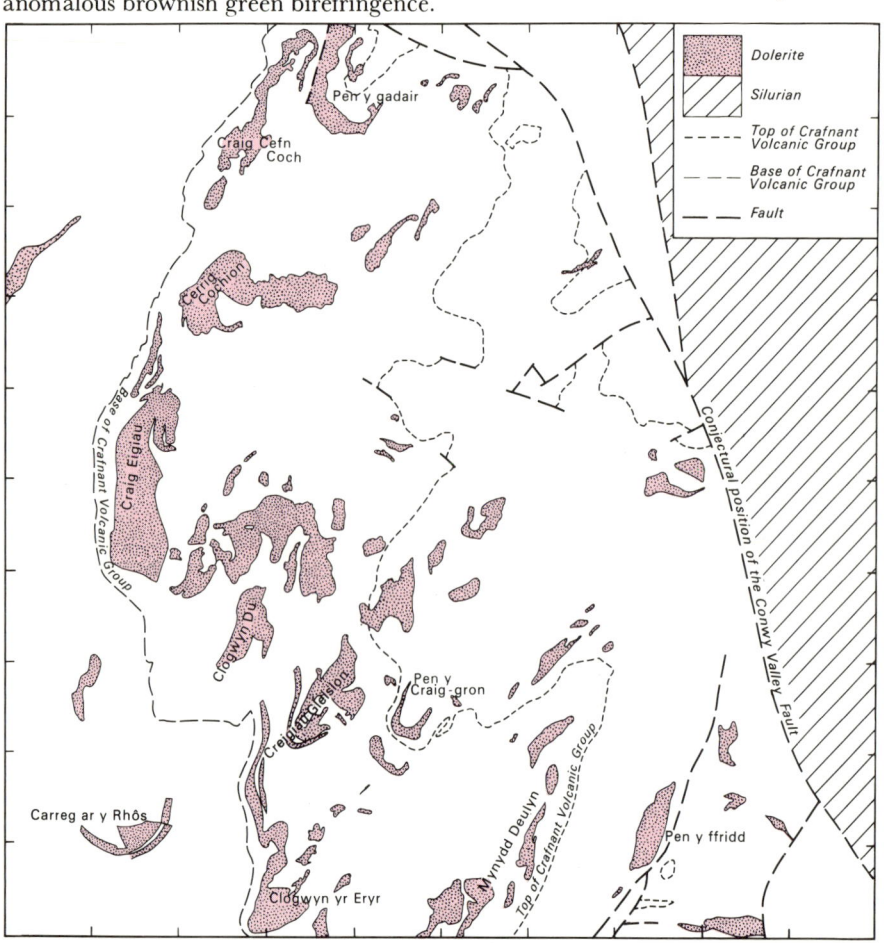

Accessories include skeletal crystals of leucoxenised titaniferous iron ore and magnetite. Leucoxene locally grades into microcrystalline sphene, clusters of which often fringe the magnetite. Small irregular grains of sulphide ore and slender prisms of apatite are common (E 40685). Epidote in prismatic sub-idiomorphic crystals, often clustered, is a common accessory, probably resulting from the albitisation of the original calcic feldspar. It is often enclosed in albitised feldspar (E 39753).

Intersertal patches and larger ocelli, filled in part by fine-grained, sheaf-like or spherulitic aggregates of feldspar (possibly oligoclase), are randomly distributed (E 40685). Amygdales are often cored by aggregates of pale green penninitic chlorite. Large areas of carbonate (E 40682) may in places be interstitial although generally such occurrences are thought to be secondary. In these specimens sheafs of acicular colourless tremolite-actinolite crystals penetrate many feldspars apparently remote from visible pyroxene and may be of primary deuteric origin.

Two occurrences, at Siglen near Llyn Cowlyd [7440 6423] and at Bwlch-y-gwynt [7762 6034], show strong spilitic affinities. In thin section (E 38924) acicular laths of albitised feldspar, serially developed and up to 1.5 mm, with intersertal chlorite form the main fabric with no indications of any original ferromagnesian minerals. Poikilitic crystals of pyrite enclose plagioclase laths (Figure 32). Carbonate-filled vesicles are often associated with irregular zones of finer feldspar laths on their peripheries and possibly represent a glassy mesostasis. The Bwlch-y-gwynt exposure shows spheroidal weathering and is clearly intrusive; at Siglen, however, there is the possibility that the spilitic rock is extrusive.

Figure 32
'Spilitic' dolerite, poikilitic pyrite with included feldspar laths, E 38924. Plane polarised light, × 40

Dolerites in the ground between Tal y fan (SH 77) and Moel Siâbod (SH 75) have been studied geochemically by Floyd and others (1976) who included material from intrusions at Pen y gadair, Cerrig Cochion, Craig Eigiau, Pen-y-ffrîdd and Craig Wen, within the present district. It was concluded that the dolerites are principally representative of continental (tholeiitic) basalts with mildly alkaline affinities developed within a stable continental plate. In addition a chemical symmetry was found to exist on either side of Cerrig Cochion and Craig Eigiau where the dolerites show low Ti, Zr and Y contents similar to ocean floor basalts.

Figure 33
Craig Eigiau and the wall of the Eigiau Dam which was breached at 8.45 pm on Monday 2nd November 1925 causing sixteen deaths in Dolgarrog village

Extent of the dolerite intrusions

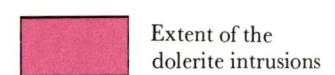

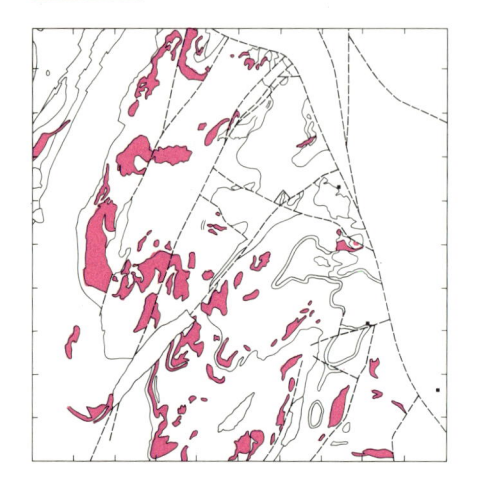

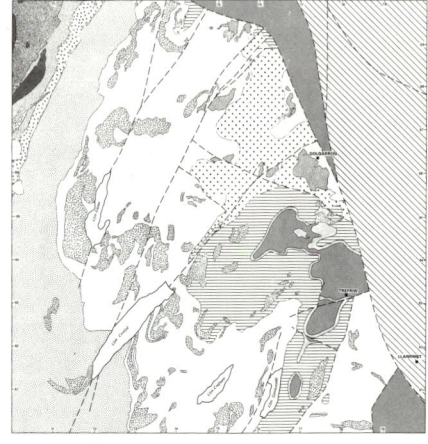

Pleistocene and Recent deposits 13

North Wales was subjected to severe glaciation during the Pleistocene period, with cold periods of glacial advance separated by warmer interglacial episodes. The knowledge of glacial chronology in the area has been summarised by Whittow and Ball (1970). Material resulting from the earlier glaciations was largely removed by the more recent, though the deposits of the latter, called variously Devensian, Würm or Weischel, are extensive.

In the district the pre-glacial land surface comprised part of the mountainous area of the Carneddau flanked on the east by a tract of lower relief, about 365 m OD, which corresponds to the 'Middle Peneplain' of Brown (1960), with some development of the 'Low Peneplain' west of Llanwrst. The remains of a higher erosion surface (the 'High Plateau') were also recognised at Pen y castell by Brown who assumed it to have a Miocene age in contrast to the probable Pliocene age of the Middle Peneplain and Low Peneplain.

During the Devensian glaciation the Conwy valley was occupied by a mass of ice which has been interpreted as a major valley glacier (Embleton, 1961; but see also Warren and others, *in press*). The movement of ice, shown by the alignment of drumlins, was to the north-north-west, along the valley.

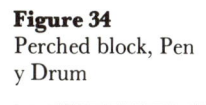

Figure 34
Perched block, Pen y Drum

Tributary glaciers occupied the valleys in the high ground to the west of the Conwy valley and here the ice movement was in a general north-easterly direction. On the eastern side of the valley there is evidence of ice spilling north-eastwards over the higher ground.

In the high ground to the west of the Conwy, glacial features include corries or cwms at Llyn Dulyn (Figure 35) and Melynllyn. Glaciated surfaces are common and roches moutonnées and perched blocks occur on the valley sides. Below about 240 m the side valleys hang above the main valley.

Boulder clay is the most widespread glacial deposit in the area. It is a grey to greyish brown clay with abundant cobbles and boulders of local rocks, the largest spreads being in the area between Pen y gadair and Afon Porth-llwyd (Figure 36) and to the south-west of Bwlch Cowlyd. The deposits in the former areas show complex terrace features which may represent ice-contact slopes marking halts in the retreat of the valley glacier in this area. Around Llanbedr-y-cennin and Tal-y-bont, boulder clay occurs both in hollows and as drumlins. A similar distribution of drift occurs on the Silurian outcrop.

Moraines dam both Llyn Dulyn and Melynllyn while an isolated example occurs a little to the north of Cefn Coch. In the south-west, Afon y Bedol cuts through another one [705623]. Morainic mounds marginal to the Conwy valley ice occur north-west of Dolgarrog and to the west of Tal-y-bont. Lateral

Figure 35
Cwm Dulyn. Glacially scoured cwm; crags in the Capel Curig Volcanic Formation in the foreground

Figure 36
Valley of Afon Dulyn viewed north-eastwards from near Melynllyn. Pen y Castell and Pen y gadair on right in middle distance; north Clwyd in far distance

moraine of a tributary valley glacier has been noted south of Llyn Geirionydd [7595 6005].

Glacial drainage channels were cut in places in the underlying rocks by meltwater from the ice. These features are present in places on the eastern side of the Conwy valley. The longest follows a sinuous line from Groesffordd [799 636] to Llanwrst. Most of the channels are dry valleys ending more or less abruptly and there is a tendency for them to follow the strike of the underlying silty mudstones. The channels are believed to have formed subglacially by water flowing under hydrostatic pressure. Hence the reversals in slope which occur in places.

Fluvioglacial sand and gravel is preserved only as a remnant terrace on the eastern side of the Conwy valley at Coed y Borthol [781 695]. It was probably deposited by meltwater during the retreat of the Conwy valley ice. At Tal-y-bont and north of Llanwrst small flats described as kame terraces by Embleton (1961, p. 58) have proved to be features in solid rock.

Head deposits, mainly gravelly clay with angular to subangular rock fragments, were formed by the effects of solifluction under periglacial conditions after the retreat of the ice. On most hill slopes they are too thin to delimit, except locally, as on the north side of Llyn Cowlyd. At Coed Dolgarrog [7695 6649] a deposit of ferrocreted boulders and coarse sand, probably derived from the solifluction of boulder clay, has been shown as head on the map.

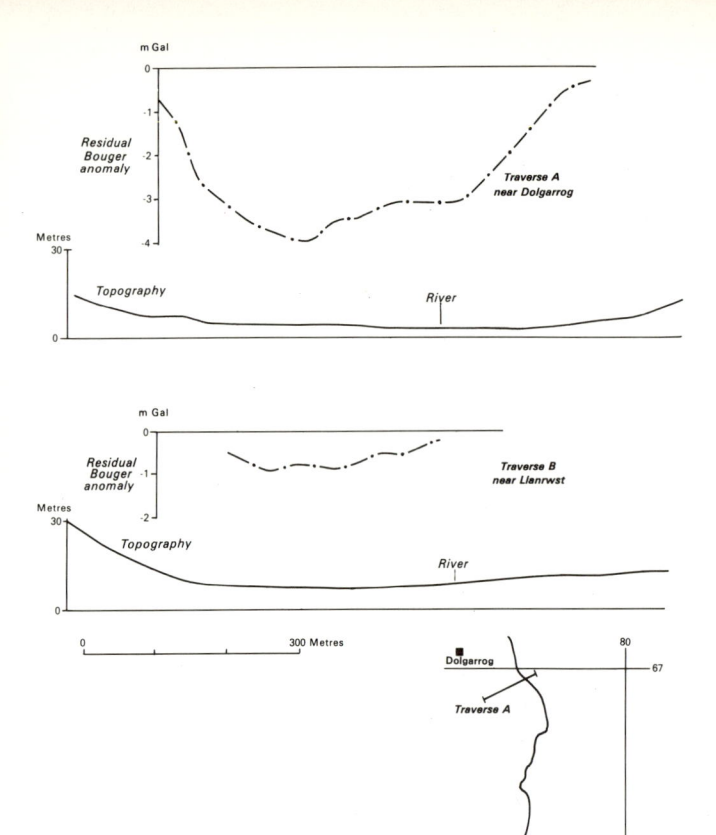

Figure 37 Geophysical (gravity) traverses over the Conwy valley alluvium

Scree forms extensive spreads on the slopes above Llyn Cowlyd. It, too, probably originated under periglacial conditions, but has continued to form up to the present day. Most of the screes are stable at their lower limits.

Lacustrine alluvium has been recognised in several places, including two areas near Llanrhychwyn. At Cwm Tal-y-braich [704 618] lacustrine clays rest on boulder clay. Such deposits are probably of late glacial origin and were formed by the temporary damming of meltwater. Thomas (1972) found traces of diatomaceous earth in the alluvium around Llyn Geirionydd. At the south-western end of Llyn Crafnant he recorded about 1.7 m of pure diatomaceous earth with thin peat layers, overlain by 1.2 to 2.4 m of peat; these deposits may extend under the lake.

The alluvium of the River Conwy is usually about 1 km wide and is liable to flooding in many places. The banks of the river show up to 4 m of grey to brown silty clay with some gravel and a

Diatoms

little peat. Geophysical (gravity) work by the Institute (Figure 37) indicates a thickness of up to 35 m of drift on bedrock at Llanwrst while around Dolgarrog the estimated thickness of drift beneath the valley floor is at least 130 m; both may include some boulder clay in the lower part.

River terraces are absent except at Llanwrst. Alluvial fans are common and of varying size, occurring both in the tributary valleys on the higher ground and where the larger tributaries join the Conwy valley.

Peat forms an extensive cover in the higher valleys expecially on the east side of Llyn Eigiau and at Pant y Griafolen. Some large patches also occur between crags on the higher ridges, for example north-east of Pen Llithrig-y-wrâch and to the east of Drum. The thickness seldom exceeds one metre, though thicker deposits are present in the south-west corner of the area. In the Conwy valley, peat overlies the alluvium in the vicinity of Maenan Abbey.

Landslips are rare though one, measuring 400 m across, has been recognised [702 632] on the south side of Cwm Eigiau (Figure 38). The beds affected are sediments of the Cwm Eigiau Formation, which have become detached from the ridge of Y Lasgallt. At Dolgarrog [769 670] a landslip has formed through movement of boulder clay on the steep side of the Conwy valley.

Figure 38
Cwm Eigiau, looking westwards to Y Lasgallt ridge with large landslip, surrounded by scree, upper centre

References

ARCHER, A. A. 1959. The distribution of non-ferrous ores in the Lower Palaeozoic rocks of North Wales. Pp. 259–276 in *Future of non-ferrous mining in Great Britain and Ireland*. (London: Institution of Mining and Metallurgy.)

BASSETT, D. A., WHITTINGTON, H. B. and WILLIAMS, A. 1966. The stratigraphy of the Bala district, Merionethshire. *Q. J. Geol. Soc. London*, Vol. 122, pp. 219–271.

BEAVON, R. V. 1963. The succession and structure east of the Glaslyn river, North Wales. *Q. J. Geol. Soc. London*, Vol. 119, pp. 479–512.

— FITCH, F. J. and RAST, N. 1961. Nomenclature and diagnostic characters of ignimbrites with reference to Snowdonia. *Liverpool Manchester Geol. J.*, Vol. 2, pp. 600–611.

BOSWELL, P. G. H. 1943. The Salopian rocks and geological structure of the country around Eglwys-fâch and Glan Conway, north-east Denbighshire. *Proc. Geol. Assoc.*, Vol. 54, pp. 93–112.

— 1949. *The Middle Silurian rocks of North Wales*. (London: Edward Arnold.)

— 1953. The alleged sub-aqueous sliding of large sheets of sediment in Silurian rocks of North Wales. *Liverpool Manchester Geol. J.*, Vol. 1, pp. 148–152.

— and DOUBLE, I. S. 1940. The geology of an area of Salopian rocks west [sic] of the Conway valley, in the neighbourhood of Llanwrst, Denbighshire. *Proc. Geol. Assoc.*, Vol. 51, pp. 151–187.

BRENCHLEY, P. J. 1964. Ordovician ignimbrites in the Berwyn Hills, North Wales. *Geol. J.*, Vol. 4, pp. 43–54.

— 1969. The relationship between Caradocian volcanicity and sedimentation in North Wales. Pp. 181–199 in *The Pre-Cambrian and Lower Palaeozoic rocks of Wales*. WOOD, A. (Editor). (Cardiff: University of Wales Press.)

BROMLEY, A. V. 1969. Acid plutonic igneous activity in the Ordovician of North Wales. Pp. 387–408 in *The Pre-Cambrian and Lower Palaeozoic rocks of Wales*. WOOD, A. (Editor). (Cardiff: University of Wales Press.)

BROWN, E. H. 1960. *The relief and drainage of Wales*. (Cardiff: University of Wales Press.)

CAVE, R. 1965. The Nod Glas sediments of Caradoc age in North Wales. *Geol. J.*, Vol. 4, pp. 279–298.

DAVIES, D. A. B. 1936. Ordovician rocks of the Trefriw district (North Wales). *Q. J. Geol. Soc. London*, Vol. 92, pp. 62–90.

DAVIES, R. A. 1969. Geological succession and structure of Cambrian and Ordovician rocks in north-eastern Carneddau. Unpublished PhD Thesis, University of Wales.

DEAN, W. T. 1958. The faunal succession in the Caradoc Series of south Shropshire. *Bull. Br. Mus. Nat. Hist.* (Geol.), Vol. 3, pp. 191–231.

DENNISON, J. B, and VARVILL, W. W. 1952. Prospecting with the diamond drill for lead-zinc ores in the British Isles. *Trans. Inst. Min. Metall. London*, Vol. 62, pp. 1–21.

DEWEY, H. and SMITH, B. 1922. Lead and zinc ores in the pre-Carboniferous rocks of West Shropshire and North Wales. Part II, North Wales. *Spec. Rep. Miner. Resour., Mem. Geol. Surv. G.B.*, Vol. 23.

DEWEY, J. F. 1969. Evolution of the Appalachian/Caledonian Orogen. *Nature, London*, Vol. 222, pp. 124–129.

DIGGENS, J. N. and ROMANO, M. 1968. The Caradoc rocks around Llyn Cowlyd, North Wales. *Geol. J.*, Vol. 6, pp. 31–48.

ELLES, G. L. 1909. The relation of the Ordovician and Silurian rocks of Conway (North Wales). *Q. J. Geol. Soc. London*, Vol. 65, pp. 169–194.

EMBLETON, C. 1961. The geomorphology of the Vale of Conway, with particular reference to its deglaciation. *Trans. Inst. Br. Geogr.*, Vol. 29, pp. 47–70.

EVANS, C. D. R. 1968. Geological succession and structure of the area east of Bethesda. Unpublished PhD Thesis, University of Wales.

FERRAR, A. M. 1973. The depth of some lakes in Snowdonia. *Geogr. J.*, Vol. 139, Part 3, pp. 516–519.

FISHER, R. V. 1966. Rocks composed of volcanic fragments and their classification. *Earth Sci. Rev.*, Vol. 1, pp. 287–298.

FISKE, R. S. and MATSUDA, T. 1964. Submarine equivalents of ash flows in the Tokiwa Formation, Japan. *Am. J. Sci.*, Vol. 262, pp. 76–106.

FITTON, J. G. and HUGHES, D. J. 1970. Volcanism and plate tectonics in the British Ordovician. *Earth Planet. Sci. Lett.*, Vol. 8, pp. 223–238.

FLOYD, P. A., LEES, G. J. and ROACH, R. A. 1976. Basic intrusions in the Ordovician of North Wales: geochemical data and tectonic setting. *Proc. Geol. Assoc.*, Vol. 87, pp. 389–400.

FRANCIS, E. H. and HOWELLS, M. F. 1973. Transgressive welded ash-flow tuffs among the Ordovician sediments of N.E. Snowdonia, N. Wales. *J. Geol. Soc. London*, Vol. 129, pp. 621–641.

GUNN, P. J. 1973. Location of the Proto-Atlantic Suture in the British Isles. *Nature, London*, Vol. 242, pp. 111–112.

HARLAND, W. B. and GAYER, R. A. 1972. The Arctic Caledonides and earlier Oceans. *Geol. Mag.*, Vol. 109, pp. 289–314.

HARKER, A. 1889. *The Bala Volcanic Series of Caernarvonshire.* (Cambridge.)

HELM, D. G., ROBERTS, B. and SIMPSON, A. 1963. Polyphase folding in the Caledonides south of the Scottish Highlands. *Nature, London*, Vol. 200, pp. 1060–1062.

HOWELLS, M. F., LEVERIDGE, B. E. and EVANS, C. D. R. 1971. The Lower Crafnant Volcanic Group, eastern Snowdonia. *Proc. Geol. Soc. London*, No. 1664, pp. 284–285.

— — — 1973. Ordovician ash-flow tuffs in eastern Snowdonia. *Rep. Inst. Geol. Sci.*, No. 73/3, 33 pp.

— FRANCIS, E. H., LEVERIDGE, B. E. and EVANS, C. D. R. 1978. *Capel Curig and Betws-y-Coed: Description of 1:25000 sheet SH 75.* Classical areas of British geology, Institute of Geological Sciences. (London: Her Majesty's Stationery Office.)

— LEVERIDGE, B. E., ADDISON, R., EVANS, C. D. R and NUTT, M. J. C. 1979. The Capel Curig Volcanic Formation, Snowdonia, North Wales; variations in ash-flow tuffs related to emplacement environment. Pp. 611–618 in *The Caledonides of the British Isles–reviewed*. HARRIS, A. L., HOLLAND, C. H. and LEAKE, B. E. (Editors).

INESON, P. R. and MITCHELL, J. G. 1975. K-Ar isotopic age determinations from some Welsh mineral localities. *Trans. Inst. Min. Metall. London, Sect. B: Appl. Earth Sci.*, Vol. 84, pp. B7–B16.

JACKSON, D. E. 1961. Stratigraphy of the Skiddaw Group between Buttermere and Mungrisdale, Cumberland. *Geol. Mag.*, Vol. 98, pp. 548–550.

JEANS, P. J. F. 1973. Plate tectonic reconstruction of the Southern Caledonides of Great Britain. *Nature, London*, Vol. 245, pp. 120–122.

JONES, O. T. 1938. On the evolution of a geosyncline (Anniversary address). *Q. J. Geol. Soc. London*, Vol. 94, pp. lx–cx.

— 1954. The use of graptolites in geological mapping. *Liverpool Manchester Geol. J.*, Vol. 1, pp. 246–260.

LAPWORTH, C. 1879. On the tripartite classification of the Lower Palaeozoic rocks. *Geol. Mag.*, Vol. 6, pp. 1–15.

MACDONALD, G. A. 1972. *Volcanoes*. (Englewood Cliffs: Prentice Hall.)

MARENGWA, B. S. I. 1973. The mineralisation of the Llanrwst area and its relation to mineral zoning in North Wales, with reference to the Halkyn-Minera area and Parys Mountain. Unpublished PhD Thesis, University of Leeds.

NUTT, M. J. C. and LEVERIDGE, B. E. *In preparation. Conwy: Description of 1:25 000 sheet SH 77*. Classical areas of British geology, Institute of Geological Sciences. (London: Her Majesty's Stationery Office.)

OLIVER, R. L. 1954. Welded tuffs in the Borrowdale Volcanic Series, English Lake District, with a note on similar rocks in Wales. *Geol. Mag.*, Vol. 91, pp. 473–483.

PETTIJOHN, F. J. 1957. *Sedimentary Rocks* (2nd Edit.) (New York: Harper.)

PHILLIPS, W. E. A., STILLMAN, C. J. and MURPHY, T. 1976. A Caledonian plate tectonic model. *J. Geol. Soc. London*, Vol. 132, pp. 579–609.

POWELL, D. W. 1971. A model for the Lower Palaeozoic evolution of the southern margin of the early Caledonides of Scotland and Ireland. *Scott. J. Geol.*, Vol. 7, pp. 369–372.

RAMSAY, A. C. 1866. The Geology of North Wales. *Mem. Geol. Surv. G.B.*, Vol. 3.

— 1881. The Geology of North Wales (2nd Edit.). *Mem. Geol. Surv. G.B.*, Vol. 3.

RAST, N. 1961. Mid-Ordovician structures in south-western Snowdonia. *Liverpool Manchester Geol. J.*, Vol. 2, pp. 645–652.

— 1969. The relationship between Ordovician structure and volcanicity in Wales. Pp. 305–335 in *The Pre-Cambrian and Lower Palaeozoic Rocks of Wales*. WOOD, A. (Editor). (Cardiff: University of Wales Press.)

— BEAVON, R. V. and FITCH, F. J. 1958. Sub-aerial volcanicity in Snowdonia. *Nature, London*, Vol. 181, p. 508.

REEDMAN, A. J., WEBB, B. C., HOWELLS, M. F., LEVERIDGE, B. E. AND BERRIDGE, N. G. *In preparation*. *Betheseda and Foel-Fras: Description of parts of 1:25000 sheets SH 66 and 67.* Classical areas of British geology, Institute of Geological Sciences. (London: Her Majesty's Stationery Office.)

RITTMANN, A. 1962. *Volcanoes and their activity.* (New York: Wiley), 305 pp.

SEDGWICK, A. 1843. Outline of the geological structures of North Wales. *Proc. Geol. Soc. London*, Vol. 4, pp. 212–224.

SHACKLETON, R. M. 1954. The structural evolution of North Wales. *Liverpool Manchester Geol. J.*, Vol. 1, pp. 261–297.

SHERLOCK, R. L. 1919. The geology and genesis of the Trefriw Pyrites deposit. *Q. J. Geol. Soc. London*, Vol. 74, pp. 106–115.

SILVESTER, N. L. 1922. The igneous rocks of Y Foel Frâs, Caernarvonshire. *Geol. Mag.*, Vol. 59, pp. 134–139.

SILVESTRI, S. C. 1963. Proposal for a genetic classification of hyaloclastites. *Bull. Volcanol.*, Vol. 25, pp. 315–322.

SMITH, R. L. 1960. Ash flows. *Bull. Geol. Soc. Am.*, Vol. 71, pp. 795–842.

STEVENSON, I. P. 1971. The Ordovician rocks of the country between Dwygyfylchi and Dolgarrog, Caernarvonshire. *Proc. Yorkshire Geol. Soc.*, Vol. 38, pp. 517–548.

TAZIEFF, H. 1974. The making of the earth. *Volcanoes and Continental drift.* (Glasgow: Robert MacLehose, University Press.)

THOMAS, D. 1972. Diatomaceous deposits in Snowdonia. *Rep. Inst. Geol. Sci.*, No. 72/5.

THOMAS, T. M. 1961. *The mineral wealth of Wales and its exploitation.* (Edinburgh and London: Oliver and Boyd.)

WARREN, P. T., PRICE, D., NUTT, M. J. C. and SMITH, E. G. *In press*. Geology of the country around Rhyl and Denbigh. *Mem. Geol. Surv. G.B.*

WHITTOW, J. B. and BALL, D. F. 1970. North-west Wales. Pp. 21–58 in *The Glaciations of Wales and adjoining regions.* LEWIS, C. A. (Editor). (London: Longmans.)

WILLIAMS, D. 1930. The geology of the country between Nant Peris and Nant Ffrancon, Snowdonia. *Q. J. Geol. Soc. London*, Vol. 86, pp. 191–233.

WILLIAMS, H. 1927. The geology of Snowdon (North Wales). *Q. J. Geol. Soc. London*, Vol. 83, pp. 346–431.

— and BULMAN, O. M. B. 1931. The geology of the Dolwyddelan Syncline (North Wales). *Q. J. Geol. Soc. London*, Vol. 87, pp. 425–458.

WILSON, J. T. 1966. Did the Atlantic close and then re-open? *Nature, London*, Vol. 211, pp. 676–681.

WOOD, D. S. and HARPER, J. C. 1962. Notes on a temporary section in the Ordovician at Conway, North Wales. *Liverpool Manchester Geol. J.*, Vol. 3, pp. 177–186.

Excursion itineraries

The district lies within the Snowdonia National Park, but most of the ground is owned privately or by the Forestry Commission. Permission should be obtained for access to any areas away from public footpaths and what is clearly open sheep grazing on the higher ground. In addition, users of the excursion itineraries set out below are strongly recommended to conform to the Code of Conduct for Geology published by the Geologists' Association.

A Lower Crafnant Volcanic Formation, south-east side of Llyn Cowlyd (½ day)

Route By car from Trefriw, turn off the Crafnant road, pass the cemetery and follow the narrow road across Cefn Cyfarwydd to the vicinity of Brwynoguchaf. Walk to the base of the Crafnant Volcanic Group [7270 6156] above the screes on the south-east side of Llyn Cowlyd.

1 *No. 1 Tuff* This lowest tuff of the Lower Crafnant Volcanic Formation shows a crude upward grading. Isolated small lithic clasts are concentrated near the base where the matrix is distinctly micaceous. Above, feldspar crystals and small pumice clasts are scattered throughout with the matrix consisting of cuspate shards in a fine-grained siliceous base. Thin impersistent silty intercalations interrupt the generally massive character of the tuff.

2 *Strata between Nos. 1 and 2 tuffs* These beds, mainly siltstone, are intruded by two dolerite sills. Near the centre of the sequence a thin acidic air-fall tuff/tuffite occurs.

3 *No. 2 Tuff* Typically uniformly fine-grained and massive. The matrix is of fine chlorite and mica in the lowest part and finely siliceous above. The tuff is shard-rich with few small crystals of feldspar and quartz. Carbonate concretions occur throughout and siliceous concretions are also present near the top.

4 *Strata between Nos. 2 and 3 tuffs* Here the beds consist entirely of cleaved black siltstones.

5 *No. 3 Tuff* The tuff is characterised by extreme coarseness and heterogeneity. The basal part is particularly rich in clasts of pumice and vitric tuff with lesser amounts of basic tuff and siltstone. Shards are extremely variable in size and form and are closely packed.

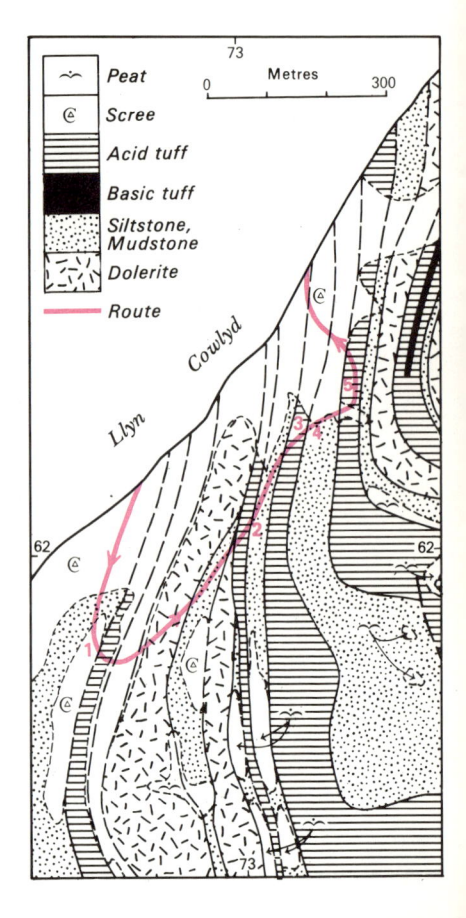

B Middle and Upper Crafnant Volcanic formations, north side of Llyn Geirionydd and Mynydd Deulyn (½ day)

Route From the main road at the north end of Llyn Geirionydd, cross the stile and take the path towards Taliesin's Monument. Continue westwards up the hill, skirting the highest crags, to the line of quarries at Clogwyn y fuwch.

1, 2 *Upper Crafnant Volcanic Formation* Small scarps in cleaved bluish grey muddy tuffite, parts rich in feldspar crystals and small clasts. Make a detour northwards to stream (2) where the tuffites are crossed by mineralised lodes, formerly worked from the Klondyke Mine.

3 *Slate at top of Middle Crafnant Volcanic Formation* Marked by a prominent slack feature with small exposures of black slate.

4 *Middle Crafnant Volcanic Formation* Dip-slope feature, on east-facing limb of anticline, in evenly bedded acid tuff-tuffite-siltstone sequence.

5 *Carbonate nodules* Small quarry in bedded tuffs on axis of anticline with cavities, up to 1.5 m in cross section, resulting from the solution of carbonate nodules.

6, 7 Cross well-defined scarps in bedded tuffs, tuffites and silty tuffs of the Middle Crafnant Volcanic Formation, accumulations of both ash-flow and ash-fall eruptions, to the slate workings at Clogwyn y fuwch (7). Here the slates have been excavated in a line of caverns, with the base of the silty crystal tuffite of the upper formation forming the hanging wall.

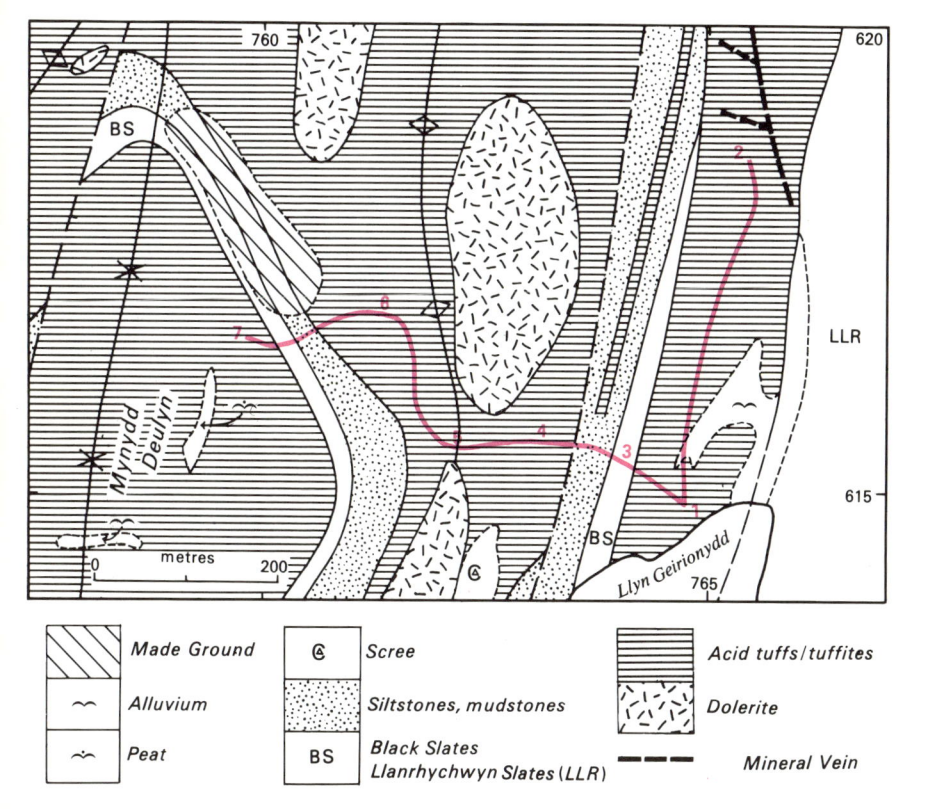

C Grinllwm Slates and Trefriw Tuff, north-north-west of Trefriw (½ day)

Route Take the Crafnant road from Trefriw, at the edge of the village bear right on the Cowlyd road, past the cemetery to Cefn Cyfarwydd. Leave the road at 1 and walk across open moorland.

1 *Llanrhychwyn Slates* Small exposure in vicinity of road consisting of evenly cleaved black slates with ochreous-weathered joint and cleavage surfaces. The slates are commonly bleached in contact with overlying peat.

2 *Trefriw Tuff* Scarps of bedded basic tuff and sandstones with a distinctive amount of basic volcanic debris. Coarser layers contain basaltic blocks up to 20 cm in length.

3 *Grinllwm Slates* Scarp features with isolated outcrops of cleaved grey silty mudstone around synclinal feature.

4, 5 *Trefriw Tuff* Bedded basic tuffs and tuffites with sandstone bands exposed in the cores of two anticlines forming prominent features at the surface, numerous exposures with sedimentary structures including cross-bedding.

6 *Dolerite* Large quarry in dolerite showing recent (1973) slip along joint surface.

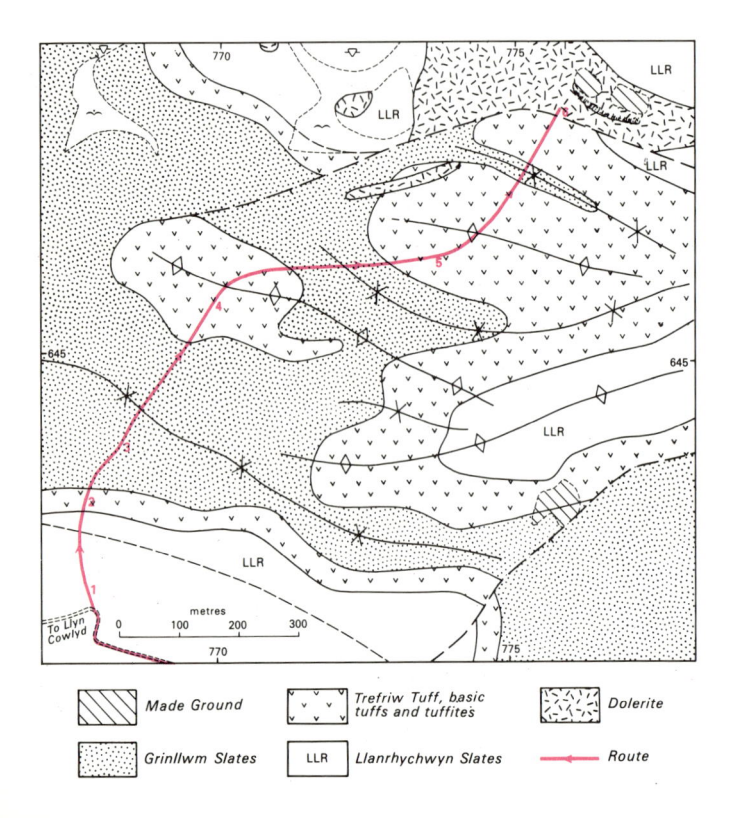

D Middle and Upper Crafnant Volcanic Formations, Dolgarrog Volcanic Formation and intrusive rhyolite, around Dolgarrog (½ day)

Route Take the Llyn Eigiau road (narrow road with passing places) from Tal-y-bont, turn off to the south to Pont Newydd [7588 6715]. Examine sections downstream from the bridge and in crags 200 m to the east.

1 *Dolgarrog Volcanic Formation* Water-washed sections, downstream from the bridge, show pillow breccias and massive hyaloclastites. Crystal-rich basic agglomeratic tuffs occur in the crags to the east.

Route From Dolgarrog, follow the path westwards along the south bank of Afon Porthllwyd to the valley scarp; follow the base of the slope southwards examining the sections.

2 *Rhyolite* Crags [7667 6761] of massive nodular rhyolite with some flow banding and cleaved in places.
Middle and Upper Crafnant Volcanic formations Dark grey tuffaceous siltstone with minor tuffites and a tuff horizon. The tuffites are pale grey, banded and reworked. The tuff, 8 m thick, forms a prominent feature and includes indurated mudstone clasts, patches of feldspar crystals and parts with diffuse banding.

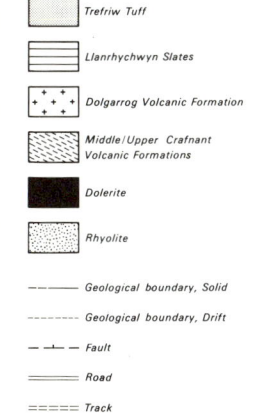

Dolgarrog Volcanic Formation Basic agglomeratic tuffs and hyaloclastites irregularly overlie the Crafnant Volcanic Group; in places faulting disrupts the contact irregularities.

Route From the Trefriw road (B5106) [7776 6562] take the Forestry Commission track to a cutting some 400 m to the north. Crags are accessible from that point and for some 600 m southwards from the upper fork of the track.

3 *Rhyolite* Well exposed showing flow banding and autobrecciation.
Dolgarrog Volcanic Formation Coarse massive basic agglomerate and fine-grained, thinly-bedded basic tuff. Towards Cae Coch the tuffs are pyritised and severely altered.

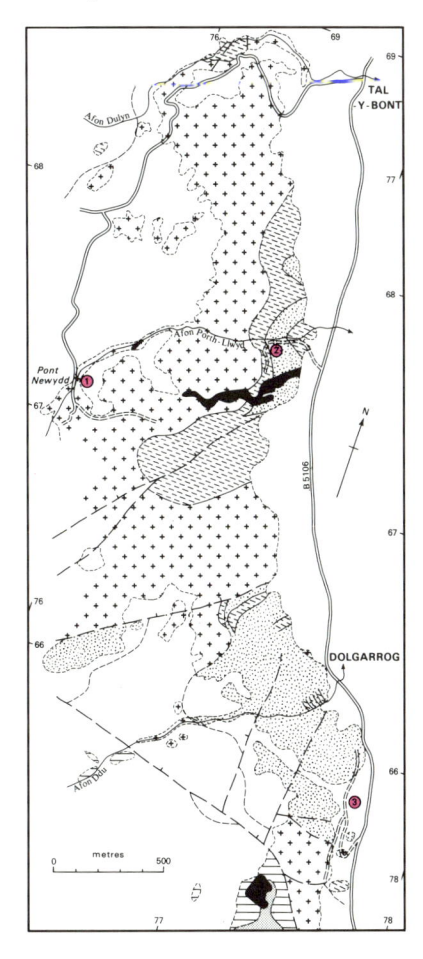

E Denbigh Grits Group (½ day)

Route From Llanrwst take Llanddoget Road to near the entrance of Plas Madoc (1), continue northwards along minor roads to the vicinity of Ty'n y celyn (2, 3) and Cefnycoedisaf (4). Return to Llanrwst along the A470 via Tan-lan (5). Car journey approximately 16 km.

1 *Old Quarry* [7986 6302] Coarsely bedded, massive sandstone, overlain by a silty mudstone disturbed bed containing large rafts of sandstone and a derived fauna of brachiopod and bivalve shell debris, crinoid columnals and corals.

2 *Crags* [7996 6591] Bedded fine-grained sandstone with silty mudstone bands. Sole structures are abundant and include longitudinal ridge, interference ripple, load and flute casts.

3 *Crag* [7980 6705] Flaggy fine-grained sandstone, laminated in part. Thin intercalations of disturbed beds occur throughout the exposure with a thicker disturbed bed near the base.

4 *Stream section* [7878 6831] Proceeding downstream, sections in striped silty mudstone exhibit a minor anticline with a complementary syncline. To the north-east of the stream section a glacial meltwater channel, trending south-west, cuts a wooded scarp.

5 *Old Quarry* [7915 6352] Flaggy siltstone overlying massive coarsely bedded sandstone with thin siltstone bands and lenses of slump conglomerate. Sole structures in the form of primary current lineations are present.

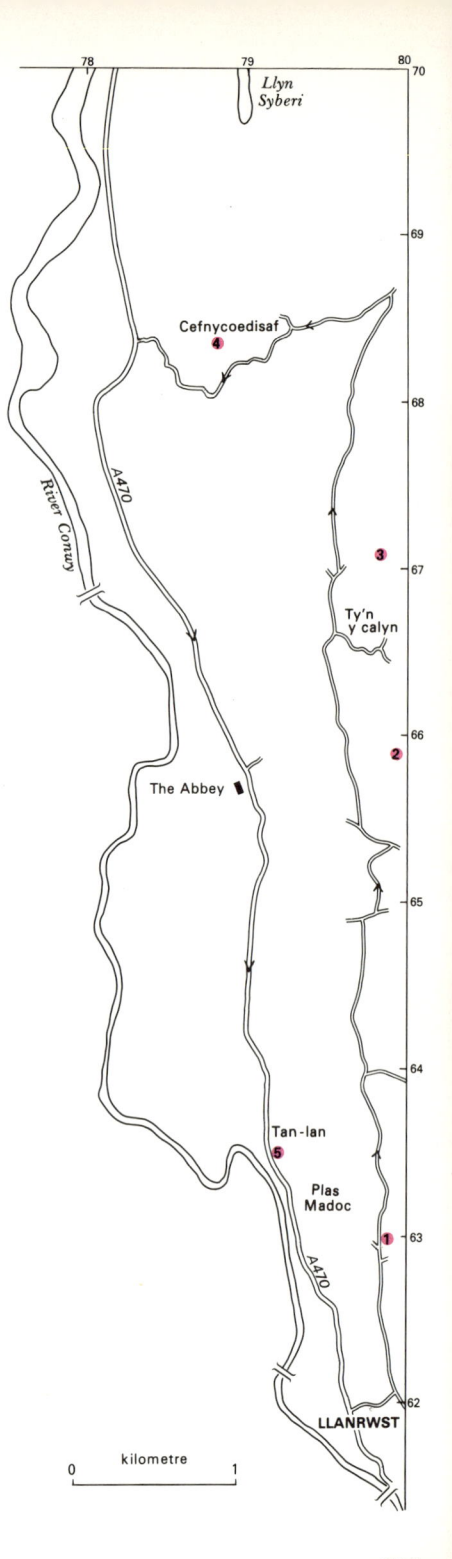

Fossil localities

The faunas associated with the Ordovician sediments are rarely well preserved and are generally deformed. Many localities have yielded restricted faunas characterised by abundant dalmanellid brachiopods with *Sowerbyella* but these, though typical of the Caradoc Series, are not indicative as to stage; diagnostic faunas are relatively few. In the Silurian also good faunas are uncommon although they include well-preserved material. A selection of the better faunas is given below. The shelly fossils were identified by Dr A. W. A. Rushton and the graptolites by Professor D. Skevington (Ordovician) and Dr R. B. Rickards and Dr P. T. Warren (Silurian).

1 [7167 6144 to 7162 6142] Crags near Adwyon Owen, south slope of Pen Llithrig-y-wrâch. Massive sandstones with abundant shells (GF 7690–7713)

Dalmanella sp. [rare], *Plaesiomys multifida* (Salter) [= *Dinorthis (Plaesiomys) multiplicata* Bancroft] [common], *Macrocoelia sp.*, *Sowerbyella sp.*, trilobite fragments.

This assemblage suggests a late Soudleyan age.

2 [7093 6542] East of honestone quarry, Melynllyn. Siltstones below the Crafnant Volcanic Group yield abundant but distorted shelly fossils suggestive of the Longvillian Stage (RU 4165–4185)

Bancroftina sp., *Dalmanella sp.*, *Sowerbyella sp.* Outcrops to the south yield *Broeggerolithus* and *Brongniartella* fragments.

3 [7180 6248] Crags, north-east side of Pen Llithrig-y-wrâch. Massive fine-grained slightly calcareous siltstone with abundant shelly fossils (GF 7747–7787)

Sponge spicules, bryozoan [stick-like colony], *Cremnorthis?*, *Dalmanella sp.*, *Dolerorthis* [fragments], *Howellites sp.*, *Kjaerina sp.*, *Reuschella?* [fragments], *Sowerbyella sp.* [common], gastropod [fragments], tentaculitids, '*Asaphus*' cf. *powisii* (Murchison), *Broeggerolithus nicholsoni*

(Reed) [common], *Brongniartella bisulcata* (McCoy), *Flexicalymene* cf. *caractaci* (Salter), primitiid ostracod [indet.], crinoid columnals [pentagonal and circular], *Orthograptus* ex gr. *calcaratus* (Lapworth).

This assemblage recalls that of the Gelli-grîn Calcareous Ashes (Longvillian) of the Bala district (Bassett and others, 1966).

4 [7745 6092 to 7760 6125] Pen-y-ffrîdd quarries (RV 6530–6550)

Amplexograptus sp., *Climacograptus brevis* Elles & Wood, *C. lineatus* Elles & Wood, *?Diplograptus multidens* Elles & Wood, *Orthograptus* ex. gr. *calcaratus*.

These represent the *Diplograptus multidens* Zone.

5 [7629 6891] Afon Dulyn, south of Ffynnon-Bedr (RV 6864–6898)

Amplexograptus sp., *Climacograptus brevis*, *C.* cf. *brevis*, *C. lineatus?*, *C. minimus?* (Carruthers), *Diplograptus* cf. *multidens*, *Glyptograptus siccatus* Elles & Wood, *Orthograptus* ex. gr. *truncatus* (Lapworth).

This assemblage represents the *Diplograptus multidens* Zone (probably the upper part).

6 [About 797 633] Plas Madoc. Greywacke sandstone with abundant fossils (Zl 9311–9362, Salter collection)

Favositid, bryozoa, *Eoplectodonta?*, *Leptaena depressa* (J. de C. Sowerby), *Meristina?*, *Microsphaeridiorhynchus nucula* (J. de C. Sowerby), *Rhynchotreta cuneata* (Dalman), *Stegerhynchus borealis* (von Buch), *S. decemplicatus* (J. de C. Sowerby), *S. diodonta* (Dalman), *Ctenodonta sp.*, *Cypricardinia sp.*, *Modiolopsis sp.*, *Nuculites* cf. *antiquus* (J. de C. Sowerby), *N.* cf. *coarctatus* (Phillips), *Cymbularia?*, *Gyronema?*, *Hormotoma?*, *Loxonema sp.*, *Platyceras haliotis* (J. de C. Sowerby), *Plectonotus trilobatus* (J. de C. Sowerby), orthocone, *Acaste sp.*, *Calymene sp.*, *Dalmanites sp.*, *Encrinurus sp.*, proetid fragments and crinoid columnals.

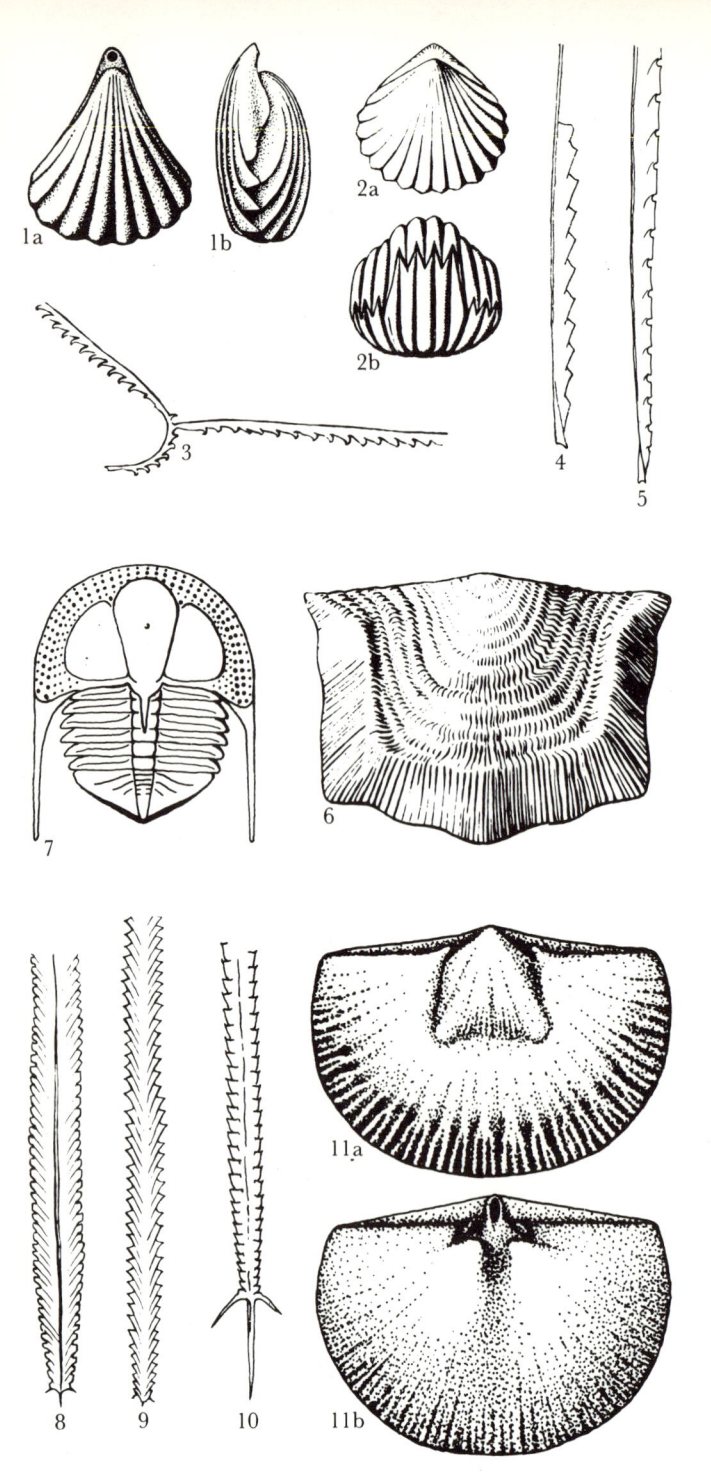

Plate illustrative of fossils from 1:25 000 sheet SH 76

Ordovician and Silurian fossils

1–6 Silurian, Wenlock Series
1a, b *Rhynchotreta cuneata* (Dalman), ×2
2a, b *Microsphaeridiorhynchus nucula* (J. de C. Sowerby), ×2
3 *Cyrtograptus rigidus* Tullberg, ×2
4 *Pristiograptus pseudodubius* (Bouček), ×4
5 *Monoclimacis flumendosae* (Gortani), ×4
6 *Leptaena depressa* (J. de C. Sowerby), ×2
7–11 Ordovician, Caradoc Series
7 *Broeggerolithus nicholsoni* (Reed), ×2
8 *Diplograptus multidens* Elles and Wood, ×2
9 *Orthograptus truncatus* (Lapworth), ×2
10 *Orthograptus calcaratus* (Lapworth), ×2
11a, b *Plaesiomys multifida* (Salter), ×2; a, ventral valve, b, dorsal valve

This locality was noted by Salter (*in* Ramsay, 1866, pp. 277–278), who also recorded a similar fauna west of Plas Madoc where it includes in addition *Atrypa reticularis* (Linnaeus) and cf. '*Phragmoceras*' *nautileum* (J. de C. Sowerby). The assemblage is indicative of the Wenlock Series.

7 [7968 6974 to 7987 6975] Stream section at Garth-iwrch (RV 3266–3286, Zp 634–640, Zp 683–695)

Cyrtograptus rigidus Tullberg, *Monoclimacis flumendosae* (Gortani), *M. sp.* aff. *vomerina* (Nicholson), *Monograptus flemingii* (Salter), *Pristiograptus pseudodubius* (Bouček).

The assemblage is diagnostic of the *Cyrtograptus rigidus* Zone.

8 [7994 6869] East-north-east of Ty'n-y-bryn (RV 3188–3215, 6925–6939)

Cyrtograptus rigidus, *Monoclimacis flumendosae*, *M. vomerina* cf. *vomerina*, *Monograptus flemingii* cf. *flemingii*, *M. flemingii* cf. var. δ.

This is a *Cyrtograptus rigidus* Zone fauna.

Glossary

Acid   Relating to igneous rocks containing more than 66 per cent of silica
Agglomerate   A volcanic rock formed of pyroclastic blocks or fragments generally more than 50 mm diameter
Albitisation   The partial or total replacement of the calcic (anorthite) component of plagioclase feldspar by sodic (albite)
Andesite   A lava of intermediate composition with andesine feldspar and one or more ferromagnesian minerals
Ash flow   A turbulent admixture of pyroclastic debris and hot gas which flows in directions imposed by the originating explosive eruption and by gravity
Axial plane   The surface that connects the axes of each plane within a fold
Basalt   A fine-grained lava or minor intrusion composed mainly of calcic plagioclase and pyroxene with or without olivine
Basic   Relating to igneous rocks containing less than 52 per cent of silica
Bioturbation   The disturbance of a sediment by organisms
Breccia   A coarse-grained clastic rock composed of angular rock fragments set in a finer-grained matrix
Caldera   A large volcanic depression, generally circular in form, which may include a vent or vents
Caledonian orogeny   Lower Palaeozoic earth movements which reached their culmination at the end of the Silurian
Columnar jointing   Prismatic fractures in lavas, sills or dykes which result from cooling
Cwm   An armchair-like hollow generally situated high on the side of a mountain; produced by the downcutting of a glacier
Devitrification   The replacement of glassy texture by crystalline texture in a volcanic rock during or after cooling
Dolerite   A medium-grained igneous rock generally forming minor intrusions and consisting mainly of calcic plagioclase and pyroxene, commonly with an ophitic texture, and sometimes olivine

Epiclastic rock A sedimentary rock formed of fragments derived by weathering and erosion of older rocks

Eutaxitic texture The texture in tuffs where shards and pumice are flattened and deformed around crystal and lithic fragments

Euhedral crystal A crystal showing its natural faces without significant modification

Fiamme Collapsed pumice fragments, commonly with ragged terminations

Flame structure Flame-shaped intrusions generally of mud grade that have been squeezed upwards into the overlying, generally coarser, layer

Gangue The uneconomic minerals of an orebody

Greywacke A poorly sorted sandstone with angular to subangular quartz and feldspar fragments and a wide range of lithic fragments set in a clayey matrix

Hyaloclastite A deposit composed of comminuted basaltic glass formed by the fragmentation of the glassy skins of basaltic pillows or by the violent eruption of basaltic material under submarine conditions

Hydrothermal alteration Alteration by or in the presence of water at high temperature

Ignimbrite The product of an ash flow (see above)

Intermediate Relating to igneous rocks transitional between acid and basic

Isocline A fold with parallel limbs

Lapilli Fragments in the range of 5 to 50 mm ejected by volcanic eruption

Load cast A sole mark composed of sediment of sand grade protruding down into finer-grade material and formed as a result of unequal loading

Lode A mineral vein in consolidated rock

Magma Molten rock generated at depth and which forms extrusive or intrusive igneous rocks on solidification; the term 'magmatic' is used here to describe intrusive and/or extrusive non-pyroclastic igneous rocks

Moraine A mound of unsorted debris deposited by a glacier (in this account ground moraine is referred to as boulder clay)

Ophitic An igneous texture where prismatic plagioclase crystals are intergrown with pyroxene crystals

Perlitic texture Small-scale arcuate cracks caused by cooling in volcanic glass

Plate tectonics Global tectonics based on an earth model characterised by a number of large lithospheric plates which move on the underlying mantle

Plunge The inclination of a fold axis

Pumice A highly vesiculated glassy lava light enough to float

Recrystallisation The formation of new crystalline mineral grains in a rock. Used here to describe tuffs and rhyolite in which the original components were glass which is assumed to have devitrified prior to recrystallisation

Rhyolite An extrusive or intrusive igneous rock of acid composition commonly porphyritic and flow banded

Roche moutonnée An elongate crag scoured by glaciation with a smooth gentle upstream side and a rough steep downstream side

Shard A glass fragment, typically found in pyroclastic rocks, having distinctive cuspate margins

Sole markings A term commonly used to describe the undersurface of a bed filling underlying sedimentary structures

Solifluction The slow, viscous downhill flow of waterlogged soil or other surface material especially in regions underlain by frozen ground

Spilite An altered basalt in which the feldspar has been albitised and the dark (mafic) minerals altered to low-temperature hydrous materials

Tuff A lithified deposit of volcanic ash

Tuffite An admixture of pyroclastic (> 25 per cent) and epiclastic (> 25 per cent) material

Vent An opening through which volcanic deposits are extruded or ejected

Vesicle A small cavity in a lava formed by included gases

Vitroclastic Texture of a pyroclastic rock composed mainly of cuspate glass fragments

Volcaniclastic Composed mainly of volcanic rock fragments

Welded tuff A pyroclastic rock in which individual particles were sufficiently plastic to agglutinate

Xenolith An inclusion in an igneous rock to which it is not genetically related

Index of geographical localities

Allt goch   35, 38
Anafon, Afon   10, 11, 12, 13, 14;
   Llyn   11
Brwynog, isaf   47; uchaf   80
Bwlch y Gwrhyd   12, 13, 14
Bwlch-y-gwynt   36, 69
Cefn Cyfarwydd   47, 49, 80, 82
Cefn Coch   22, 72; Craig   33
Cefnycoedisaf   58, 84
Cerrig Cochion   29, 67, 70
Clogwyn Gwlyb   63; Coed   34, 35, 36
Clogwyn-y-Fuwch Quarry   34, 81
Clogwyn yr Eryr   28, 29
Coed Creigiau   49, 52
Cowlyd, Bwlch   21, 67, 72; Llyn   1, 2,
   20, 21, 24, 26, 28, 29, 30, 31, 32, 33, 39,
   41, 60, 64, 69, 73, 74, 80; valley   1, 59
Crafnant, Afon   1, 34, 38, 49; Llyn   1,
   24, 26, 29, 31, 33, 37, 67, 74;
   valley   1, 28, 34, 35, 37, 47, 49, 57, 59,
   61, 67
Craig Ffynnon   31, 39
Craig-wen   26, 28
Creigiau Gleision   31, 32, 38
Cwm Bychan   11, 14, 16
Cynllwyd-bach   37, 38

Ddu, Afon   1, 16, 48
Dolgarrog   1, 41, 42, 47, 60, 72, 75, 83;
   Coed   40, 65, 73
Dulyn, Afon   1, 39, 40, 41, 45, 48, 73, 85;
   Cwm   72; Llyn   1, 16, 17, 72
Dyto   54

Eigiau, Cefn Tal-llyn   27, 29; Craig   27,
   30, 67, 70; Cwm   5, 21, 28, 29, 59, 75;
   Llyn   1, 30, 60, 67, 75
Eilio   32; Moel   39, 43

Ffynnon-Bedr   48, 85
Foel Lŵyd   15, 16, 17, 18, 22
Ffridd-uchaf   49, 52

Gallt Cedryn   22, 27, 28
Garth-iwrch   87
Geirionydd, Afon   37, 63; Llyn   1, 34,
   35, 47, 73, 74, 81; valley   61
Gellilydan   37; mine   61, 62
Grinllwm   49, 51

Llanbedr-y-cennin   48, 49, 52, 60, 65, 72
Llanrhychwyn   46, 47, 59, 67, 74
Llanrwst   1, 34, 35, 52, 55, 56, 60, 61, 71,
   73, 75, 84
Maenan, Abbey   75; Plas   54
Melynllyn   1, 21, 72, 85
Mine, Cae Coch   2, 42, 48, 49, 61, 62, 64,
   67, 83
—, Coed Gwydir   61
—, Gellinewydd   61, 64
—, Glan Geirionydd   34, 61, 62
—, Hafna   36, 61, 62, 63, 67
—, Klondyke   61, 62, 63, 81
—, Llyn Geirionydd   37, 61, 62
—, Pandora   61, 62, 63
—, Parc   61, 62, 63, 64
—, Rhuad y rhwst   36, 61, 62, 63
—, Tal y llyn   61
—, Tan-yr-Eglwys   61, 62
—, Ty'n Twll   61, 62, 63
Mynydd Deulyn   37, 59, 81

Nant Gwydir   1; valley   34, 61, 63, 67

Pant y Carw mine   62; quarry   36
Penardda   42, 43, 65
Pen Llithrig-y-wrâch   5, 20, 21, 27, 75,
   85
Pen y Castell   23, 27, 28, 29, 31, 33
Pen y Drum   47, 51, 71
Pen-y-ffridd   47, 70; quarries   85
Pen y gadair   29, 30, 32, 40, 45, 67, 70, 72
Plas Madoc   84, 85; Lodge   54
Pont Newydd   83
Porth-llwyd   65, 66; Afon   1, 39, 41, 43

Rhibo   50
Rowlyn, isaf   66; uchaf   40

Sarnau   34, 39
Siglen   39, 69

Tal-y-bont   1, 45, 52, 72, 73, 83
Tal-y-braich, Cwm   20, 74; isaf   20
Tan-lan   54, 84
Trefriw   1, 49, 52, 61, 80, 82
Ty'n-y-bryn   55, 87

Y Lasgallt   5, 20, 75

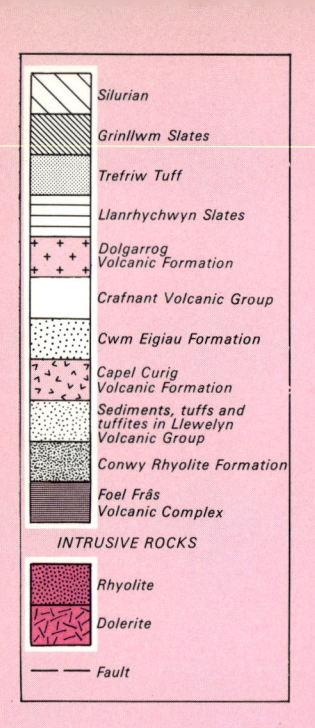

▨	Silurian
▬	Grinllwm Slates
░	Trefriw Tuff
☰	Llanrhychwyn Slates
+++	Dolgarrog Volcanic Formation
□	Crafnant Volcanic Group
░	Cwm Eigiau Formation
⌄⌄	Capel Curig Volcanic Formation
░	Sediments, tuffs and tuffites in Llewelyn Volcanic Group
▨	Conwy Rhyolite Formation
■	Foel Frâs Volcanic Complex

INTRUSIVE ROCKS

▨	Rhyolite
▨	Dolerite
---	Fault

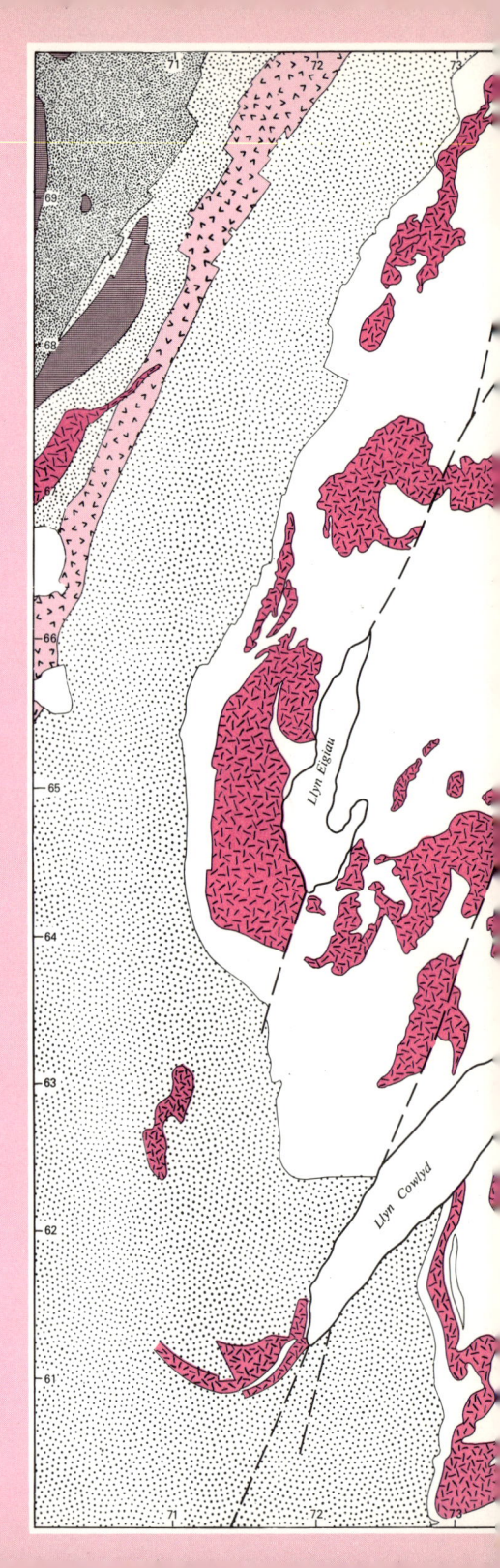